Other Titles in This Series

(Continued in the back of this publication)

AMS SHORT COURSE LECTURE NOTES
Introductory Survey Lectures
published as a subseries of
Proceedings of Symposia in Applied Mathematics

Proceedings of Symposia in
APPLIED MATHEMATICS

Volume 49

Complex Dynamical Systems
The Mathematics Behind the Mandelbrot and Julia Sets

Robert L. Devaney, Editor

Bodil Branner
Linda Keen
Adrien Douady
Paul Blanchard
John H. Hubbard
Dierk Schleicher
Robert L. Devaney

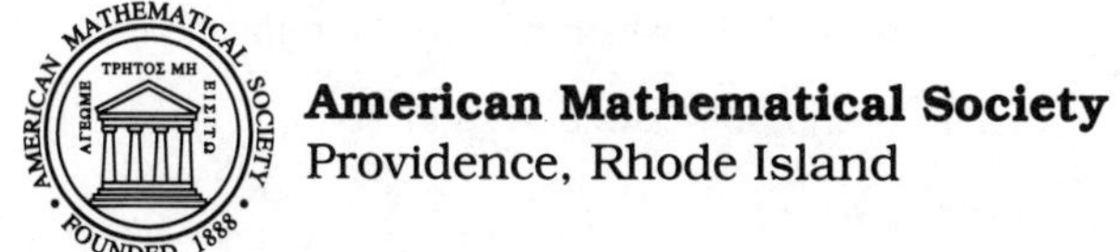

American Mathematical Society
Providence, Rhode Island

LECTURE NOTES PREPARED FOR THE
AMERICAN MATHEMATICAL SOCIETY SHORT COURSE
COMPLEX DYNAMICAL SYSTEMS

HELD IN CINCINNATI, OHIO
JANUARY 10–11, 1994

The AMS Short Course Series is sponsored by the Society's Program
Committee on National Meetings. The Series is under the direction of the
Short Course Subcommittee of the Program Committee
for National Meetings.

1991 *Mathematics Subject Classification.* Primary 58Fxx, 30Cxx.

Library of Congress Cataloging-in-Publication Data

Complex dynamical systems: the mathematics behind the Mandelbrot and Julia sets / Robert L. Devaney, editor.
 p. cm. — (Proceedings of symposia in applied mathematics, ISSN 0160-7634; v. 49)
 Includes bibliographical references and index.
 ISBN 0-8218-0290-9 (alk. paper).
 1. Differentiable dynamical systems—Congresses.   I. Devaney, Robert L., 1948– .   II. Series.
QA614.8.C645   1994
514′.74—dc20   94-24190
                                                                                                    CIP

Contents

Preface

In the last fifteen years, the Mandelbrot set has emerged as one of the most recognizable objects in mathematics. While there is no question that this set is an object of great beauty, relatively few people appreciate the fact that the mathematics behind these images is equally beautiful. The goal of the Short Course at which these lectures were given was to remedy this.

The course was called *Complex Dynamical Systems: The Mathematics Behind the Mandelbrot and Julia Sets* and was held at the annual meeting of the American Mathematical Society in Cincinnati, Ohio, on January 10–11, 1994. The lectures covered a wide range of topics in complex dynamics. Among other topics, the lectures surveyed classical work of Julia and Fatou on local dynamics of anlaytic maps as well as more recent work on the dynamics of quadratic and cubic polynomials, the geometry of Julia sets, and the structure of various parameter spaces. The lectures contain reports on recent work on Yoccoz puzzles and tableaux, limiting dynamics near parabolic points, and the spider algorithm, as well as discussions of extensions of the theory to rational maps, Newton's method, and entire transcendental functions. Much of this book should be accessible to anyone with a background in the basics of dynamical systems and complex analysis.

Robert L. Devaney

Proceedings of Symposia in Applied Mathematics
Volume **49**, 1994

The Complex Dynamics of
Quadratic Polynomials

Robert L. Devaney*

The field of complex analytic dynamics has undergone rapid development in the past fifteen years. After a period of relative dormancy, the field was rejuvenated in 1980 thanks to some intriguing computer graphics images of Mandelbrot as well as major new mathematical advances due to Douady, Hubbard, Sullivan and others.

Our goal in this paper is twofold. First, we wish to introduce the basic idea of complex analytic dynamics in the simplest possible setting, namely iteration of quadratic polynomials. Second, we hope to provide a framework which ties together the other papers in this volume.

1. Historical Overview. The field of complex analytic dynamics has experienced two relatively short periods of vigorous growth. The field traces its origins to the late nineteenth and early twentieth centuries. At that time, mathematicians such as Leau, Schröder, Kœnigs, Böttcher, among others, became interested in the local behavior of complex functions under iteration. Their work focused primarily on

---

* 1980 *Mathematics Subject Classification (1985 Revision)*. Primary 58F13; Secondary 58F14. Partially supported by National Science Foundation Grant DMS-93-02986

the behavior of such iterations near a fixed point. A good reference
for this older work is [Al].

During the period 1918-20, a dramatic shift in emphasis occurred
due primarily to the pioneering efforts of two French mathematicians,
Gaston Julia and Pierre Fatou. Instead of considering only local dynamical behavior, Julia and Fatou took a more global point of view.
Away from the fixed points, they found quite different dynamical
behavior. Sometimes the results of iteration were quite tame or stable; at other times these iterations behaved in dramatically different
fashion—what we now call chaotic behavior. In honor of the contri-
butions of these two men, we now call the stable regime for a complex
dynamical system the Fatou set, while the chaotic region is known as
the Julia set. In a series of remarkable papers in the 1920's, Fatou
and Julia succeeded in describing many of the properties of both of
these sets for rational maps. However, in attempting to classify all
possible dynamics on the Fatou set, they reached an impasse. They
could not rule out the possibility that the dynamics included wandering domains, which they expected would not occur, nor could they
prove the existence of what are now known as Siegel disks, which they
expected would occur. With these two roadbloacks in the way, work
in complex dynamics slowed, and there was not much activity in the
field for fifty years.

There were two notable events during this period of dormancy. In
the 1940's, C.L. Siegel showed that Siegel disks could in fact exist in
complex dynamical systems. This brought the classification of stable
regions one step closer to completion. Later, I.N. Baker extended
much of the work of Fatou and Julia to other classes of functions,
showing along the way that other types of stable behavior could occur
for entire and meromorphic functions.

The second major period of activity began in 1980 after Mandelbrot first used computer graphics to explore complex dynamics. His
discovery of the Mandelbrot set prompted many mathematicians to
reinvestigate this field. In quick succession, Dennis Sullivan introduced the use of quasi-conformal mappings into the subject. This
allowed him to prove his No Wandering Domains theorem, which essentially completed the classification of stable dynamics for rational
maps begun by Fatou and Julia. At the same time, Douady and
Hubbard opened new vistas in the field by considering the parameter
space for quadratic polynomials. They developed a technique which
enabled them to classify more or less completely all possible types of
quadratic dynamics. (At this time, only one obstacle remains—the

question of the local connectivity of the Mandelbrot set.) Immediately thereafter, work began on a variety of other types of complex dynamical systems, including rational maps and higher degree polynomials, systems arising from Newton's method, and entire and meromorphic functions. Many of these later developments are surveyed in subsequent chapters of this volume.

It is a pleasure to thank Tom Scavo and Dierk Schleicher for helpful comments on an earlier draft of this paper.

2. Basic Properties.

Throughout this paper we will restrict attention to the dynamics of the family of complex quadratic polynomials,

$$P_c(z) = z^2 + c.$$

Here c is a complex parameter. Our goal is to understand the behavior of this function under iteration. Given a seed $z_0 \in \mathbf{C}$, the *orbit* of z_0 under P_c is the sequence of points $z_0, z_1, z_2, \ldots$ where $z_n = P_c(z_{n-1})$ for $n = 1, 2, \ldots$. Note that $z_n = P_c^n(z_0)$ where

$$P_c^n = \underbrace{P_c \circ \ldots \circ P_c}_{n \ times}.$$

Certain orbits play a crucial role in the dynamics of P_c. These include the *fixed points* $(P_c(z_0) = z_0)$ and the *periodic points* $(P_c^n(z_0) = z_0)$. The least positive integer n for which $P_c^n(z_0) = z_0$ is called the *period* of the periodic orbit. The orbit of z_0 is *preperiodic* if it is not periodic, but $P_c^{n+j}(z_0) = P_c^j(z_0)$ for some $j > 0$. As we will see, the most important orbit of P_c is the orbit of 0. Note that $P_c'(0) = 0$. For this reason, the orbit of 0 under P_c is called the *critical orbit*. Note that 0 is the only critical point of P_c in $\mathbf{C}$.

Definition. Suppose z_0 is a fixed point of P_c. Then z_0 is

1. *attracting* if $0 < |P_c'(z_0)| < 1$,
2. *superattracting* if $P_c'(z_0) = 0$,
3. *repelling* if $|P_c'(z_0)| > 1$,
4. *neutral* if $|P_c'(z_0)| = 1$.

Periodic points of period n are similarly classified by replacing P_c with P_c^n in the above definition.

The reason for this terminology is the following. If z_0 is an attracting or superattracting fixed point, then there is an open neighborhood

U of z_0 having the property that $P_c^n(z) \to z_0$ as $n \to \infty$ for each z in U. This follows immediately from the Contraction Mapping Principle. The set of all points whose orbits converge to z_0 is called the basin of attraction of z_0.

If z_0 is a repelling fixed point, the nearby dynamical behavior is quite different. Since $P_c'(z_0) \neq 0$, it follows from the Inverse Function Theorem that there is a neighborhood U of z_0 in which there exists an analytic branch of the inverse of P_c. Since z_0 is an attracting fixed point for this inverse, it follows that orbits in U are attracted to z_0 under iteration of this inverse. Consequently, orbits of points in U are repelled from z_0 by P_c.

The intermediate case is the neutral case. Here we may have nearby zones in which we have attraction or repulsion or other types of dynamics. As is discussed in the article of Keen, the dynamics near neutral fixed points may be quite complicated and, in some cases, still not completely understood.

It is often useful to view the dynamics of P_c on the Riemann sphere $\overline{\mathbf{C}} = \mathbf{C} \cup \{\infty\}$. When $\mathbf{C}$ is compactified in this manner, the point at ∞ becomes a fixed point for P_c and moreover, $P_c'(\infty) = 0$. This can be seen by making the change of coordinates $h(z) = 1/z$ which takes ∞ to 0. In these new coordinates, P_c assumes the form

$$z \mapsto \frac{z^2}{1 + cz^2}.$$

It is easy to check that 0 is a superattracting fixed point for this map. As a consequence, there exists $R > 0$ such that, if $|z| > R$, $|P_c^n(z)| \to \infty$ as $n \to \infty$.

One of the great beauties of complex dynamical systems is the sharp division between regions where we have chaotic dynamics and sets on which the dynamics are stable. To explain this, we need to introduce several auxiliary notions.

Definition. Let Λ be a metric space and suppose $F: \Lambda \to \Lambda$ is continuous. We say that F exhibits *sensitive dependence on initial conditions* if there exists a constant $\beta > 0$ such that for any $p \in \Lambda$ and any neighborhood U of p, there exists $n > 0$ and $y \in U$ such that

$$d[F^n(p),\ F^n(y)] > \beta.$$

The basic idea behind this definition is the following. No matter how small a neighborhood U we choose about p, we may always find a point in U whose orbit at some point diverges from that of p by at least β units. As a consequence, if a dynamical system exhibits sensitive dependence, numerical computations of orbits must always be suspect, since small errors in the location of the initial seed or in the computation of the orbit may be magnified under iteration.

We say that $F\colon \Lambda \to \Lambda$ is *chaotic* if F depends sensitively on initial conditions at each point in Λ. At the other extreme, F is *stable* on Λ if it does not depend sensitively on initial conditions at any point in Λ.

Note that, in order to speak about chaotic behavior, we need the presence of a metric. In this paper, we will always assume that the metric on $\overline{\mathbf{C}}$ is bounded at ∞, so that orbits which are attracted to ∞ do so with their mutual distances apart tending to zero.

To derive a criterion for chaotic behavior, we need one additional notion.

Definition. A continuous map $F\colon \Lambda \to \Lambda$ is *transitive* if, for any open sets U and V, there exists $n > 0$ such that $F^n(U) \cap V \neq \emptyset$.

It is known that if Λ is a metric space and F is transitive, then there is an orbit under F which is dense in Λ. For more details on these definitions we refer to [D, chapter 6].

A criterion for chaos is then:

Theorem. *Suppose Λ is an infinite set. Suppose also that $F\colon \Lambda \to \Lambda$ satisfies*

    1. Periodic points for F are dense in Λ,
    2. F is transitive on Λ.

Then F is chaotic on Λ.

A proof of this result may be found in [BBCDS].

Now we come to the most important object in the study of the dynamics of quadratic functions, the Julia set.

Definition. The *Julia set* of P_c, denoted $J(P_c)$, is the set of all points at which P_c exhibits sensitive dependence. That is, $J(P_c)$ is the chaotic set for P_c. The complement of $J(P_c)$ is the Fatou set or stable set for P_c.

We will describe several important examples of Julia sets in the next section. Here we list several important alternative characterizations of the Julia set of P_c.

The classical definition of the Julia set (and the one that is most useful for proving theorems) involves the concept of normal families of functions. A collection of functions $\{F_i \,|\, i = 1, 2, \ldots\}$ is said to be *normal* (in the sense of Montel) on an open set $U \subset \mathbf{C}$ if the F_i form an equicontinuous family on U. If the F_i are holomorphic functions then normality of a family $\{F_i\}$ on U is equivalent to the following: Every sequence of the F_i has a subsequence that converges uniformly on compact sets either to an analytic function on U or to ∞. Conversely, if the $\{F_i\}$ do not form a normal family on U, then we have the important result of Montel:

Theorem. *Suppose the family of analytic functions $\{F_i\}$ is not normal on U. Then the family $\{F_i\}$ assumes all values in $\overline{\mathbf{C}}$ except at most two.*

Note that, if the family of iterates of P_c, $\{P_c^n\}$, does not form a normal family in any neighborhood of z_0, then z_0 certainly lies in $J(P_c)$. Indeed, close to z_0 there are points whose orbits tend arbitrarily far from the orbit of z_0. The converse of this result is also true, yielding the first alternative characterization of $J(P_c)$:

Theorem. $J(P_c) = \{z_0 \in \mathbf{C} \,|\, \{P_c^n\}$ *is not a normal family in any neighborhood of $z_0\}$.*

Recall that a necessary condition for a function to be chaotic on a set is the density of periodic points in that set. For analytic maps, this condition is also sufficient. Indeed, if z_0 lies on a repelling periodic orbit, then $|(P_c^n)'(z_0)| = \lambda > 1$. Hence,

$$|(P_c^{nj})'(z_0)| = \lambda^j \to \infty$$

as $j \to \infty$. Thus, no subsequence of the family $\{P_c^{nj}\}$ converges to an analytic function. Since $P_c^{nj}(z_0) = z_0$ for each j, no subsequence can converge to ∞. Thus the family $\{P_c^n\}$ is not normal in any neighborhood of z_0, and so $z_0 \in J(P_c)$.

It follows similarly that any accumulation point of repelling periodic points must also lie in $J(P_c)$, so we have another characterization of this set.

Theorem. $J(P_c)$ *is the closure of the set of repelling periodic points of P_c.*

Recall that ∞ is always an attracting fixed point for P_c. Consequently, there exists $R > 0$ such that if $|z| > R$, then $|P_c^n(z)| \to \infty$ as $n \to \infty$. Hence any such z does not lie in $J(P_c)$, and so $J(P_c)$ is a bounded set. Moreover, the orbit of any point in $J(P_c)$ is also bounded.

Definition. The *filled Julia set* of P_c, denoted $K(P_c)$, is the set of points whose orbits are bounded under iteration of P_c.

Thus $J(P_c) \subset K(P_c)$. However, since $\{P_c^n\}$ is not normal in any neighborhood of a point in $J(P_c)$, it follows that there are points arbitrarily close to any point in $J(P_c)$ whose orbit escapes to ∞. Conversely, if we have a point whose orbit is bounded, but arbitrarily nearby we have escape orbits, then by Montel's Theorem, this point must be in $J(P_c)$. Thus we have a fourth characterization of J:

Theorem. $J(P_c)$ *is the boundary of* $K(P_c)$, *i.e., the boundary between bounded orbits and orbits that escape.*

We caution the reader that this final characterization of the Julia set holds only for polynomials. For rational maps, the point at ∞ need not be superattracting, and for entire and meromorphic functions, the point at ∞ is an essential singularity. In these cases, this last characterization of the Julia set no longer holds in general. However, the first three characterizations hold for general analytic maps.

Which points lie in the Fatou set? Certainly any attracting periodic point together with its basin of attraction lies in the Fatou set. Clearly there are no repelling periodic points in a basin of attraction, nor are there escape orbits. These are the principal types of points in the stable region of P_c, but there are other types of points. We refer to the article of Keen in this volume for a more complete description of the stable set.

A natural question is why restrict to polynomials of the form $P_c(z) = z^2 + c$, rather than the general quadratic. The reason for this is that any quadratic map is dynamically equivalent to one in the family P_c. To be precise, we say that two maps $F, G \colon \overline{\mathbf{C}} \to \overline{\mathbf{C}}$ are *conjugate* if there exists a homeomorphism $h \colon \overline{\mathbf{C}} \to \overline{\mathbf{C}}$ such that $h \circ F = G \circ h$. It follows that $h \circ F^n = G^n \circ h$, so that h maps orbits of F to those of G. Similarly, h^{-1} takes orbits of G to those of F. So the conjugacy h gives a one-to-one correspondence between orbits of F and G. In the quadratic case, we have

Proposition. *Suppose $F(z) = \alpha z^2 + \beta z + \gamma$ with $\alpha \neq 0, \beta, \gamma \in \mathbf{C}$. Then F is conjugate to $P_c(z) = z^2 + c$ for some $c \in \mathbf{C}$.*

In fact, the affine conjugacy

$$H(z) = \alpha z + \beta/2$$

satisfies $H \circ F = P_c \circ H$, where

$$c = \alpha\gamma + \frac{\beta}{2} - \frac{\beta^2}{4}.$$

3. Examples of Julia sets.

In this section we briefly describe the Julia sets of several important quadratic functions. While these examples are in a certain sense special, we will see in later sections that they provide basic models for all quadratic polynomials.

Example 1. $P_0(z) = z^2$. The squaring function is the simplest quadratic function. Note that 0 and ∞ are superattracting fixed points for this map. We also have

1. $|P_0^n(z_0)| \to 0$ if $|z_0| < 1$.
2. $|P_0^n(z_0)| \to \infty$ if $|z_0| > 1$.
3. $|P_0^n(z_0)| = 1$ if $|z_0| = 1$.

Thus $\{z \mid |z| < 1\}$ is the basin of attraction for the fixed point at 0, and so belongs to the Fatou set. All points in $\{z \mid |z| > 1\}$ have orbits that escape to ∞, so these points also belong to the Fatou set. The region in between is the unit circle, which is the boundary of $K(P_0)$, so this is the Julia set of P_0.

The dynamics on $J(P_0)$ are chaotic, but nevertheless easy to describe. Let us regard the unit circle as $\mathbf{R}/\mathbf{Z}$, so that points on the circle are defined mod 1. Then the orbit of any point on $J(P_0)$ is obtained by iterating the "doubling function," $D(\theta) = 2\theta$ mod 1. It is easy to check that if $\theta = p/q$ in lowest terms with q odd, then θ lies on a periodic orbit. Consequently, periodic points are dense in the unit circle. Also, D expands arclengths on the circle by a factor of 2. Therefore any small arc is eventually mapped onto the entire circle by D, thus proving transitivity. Hence, D is chaotic. Sensitive dependence is also clear, for if we take any open neighborhood U of a point on the unit circle, and repeatedly apply P_0, we see that $P_0^n(U)$ grows in size so that $\bigcup_{n\geq 0} P_0^n(U)$ contains any point in $\overline{\mathbf{C}}$ except (possibly) 0 and ∞. Finally, if $\theta = p/(2^n q)$ where q is odd, then θ is eventually periodic under D, and so these types of points are also dense in $J(P_0)$.

Example 2. $P_{-2}(z) = z^2 - 2$. For this map the orbit of 0 is eventually fixed, since $P_{-2}^2(0) = 2$, which is a fixed point. Although this map is apparently quite different from the squaring map P_0, there is nevertheless a striking relationship. Consider the function $H(z) = z + 1/z$ defined on $\{z \mid |z| \geq 1\}$. It is easy to check that H maps the exterior of the open unit disk onto the entire plane, with the unit circle mapped in 2-to-1 fashion (except at ± 1) onto the interval $[-2, 2]$. In fact, H takes straight rays emanating from the origin onto pieces of hyperbolas as shown in Figure 1.

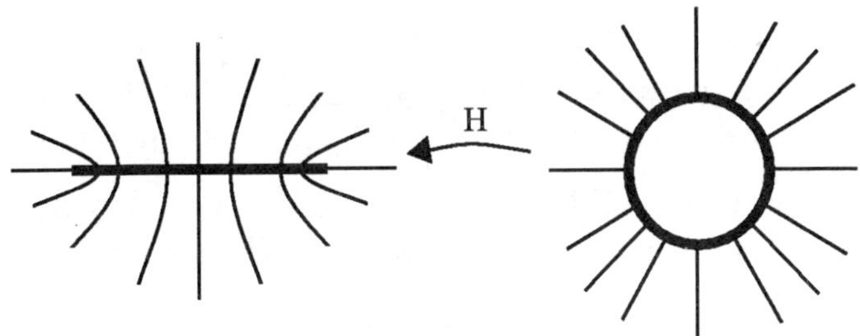

Fig. 1 The map H takes rays to pieces of hyperbolas.

More importantly, H conjugates P_0 on $\{z \mid |z| \geq 1\}$ to P_{-2}. Indeed, one computes easily that

$$H(z^2) = (H(z))^2 - 2,$$

which shows that $H \circ P_0 = P_{-2} \circ H$.

It follows that, if $z \notin [-2, 2]$, then $P_{-2}^n(z_0) \to \infty$. On the other hand, any point in $[-2, 2]$ has bounded orbit under P_{-2}. Hence, $K(P_{-2}) = [-2, 2] = J(P_{-2})$.

The mapping H is not quite one-to-one on the unit circle. Nevertheless, H does collapse the dynamics of P_0 on the unit circle to that of P_{-2} on $[-2, 2]$. Moreover, because of the conjugacy H, the dynamics of P_{-2} on the hyperbolas in the exterior of $[-2, 2]$ is precisely the same as the dynamics of P_0 on the straight rays in the exterior of the unit circle. This is an important idea: We will see that the same thing happens for every quadratic polynomial and that this allows us to analyze the dynamics of P_c on its Julia set in rather complete fashion.

In the previous two examples, we were able to compute $J(P_c)$ explicitly. Unfortunately, these are the only two polynomials of the form $z \mapsto z^2 + c$ for which this is true. For other values of c, we must resort to numerical techniques to paint the picture of the Julia set (or filled Julia set). One of the easiest algorithms to find $K(P_c)$ is the following:

Escape-time algorithm:

1. Select a grid of points in $\mathbf{C}$.
2. For each grid point z_0, compute $P_c^i(z_0)$ for $1 \le i \le N$, for some predetermined value of N.
3. If $|P_c^i(z_0)| \ge$ BOUND for some i, then assume the orbit of z_0 escapes and color z_0 white (not in $K(P_c)$).
4. If $|P_c^i(z_0)| <$ BOUND for all $i \le N$, then assume that the orbit of z_0 is bounded and color z_0 black (in $K(P_c)$).

The precise value of BOUND in the above algorithm is max $(2, |c|)$. This can be seen as follows. Let $B = \max(2, |c|)$. Then, if $|z| > B$,

$$\frac{|P_c(z)|}{|z|} \ge |z| - \frac{|c|}{|z|} > |z| - 1 > 1.$$

Thus, if $|z| > B$, it follows that

$$|P_c^{n+1}(z)| > |P_c^n(z)|$$

and so $|P_c^n(z)| \to \infty$.

Example 3. $P_{-1}(z) = z^2 - 1$. For this polynomial, $P_{-1}(0) = -1$, $P_{-1}(-1) = 0$, and so 0 and -1 lie on a periodic orbit of period 2. Also $P_c'(0) = 0$ for all c, and so

$$(P_{-1}^2)'(0) = (P_{-1}^2)'(-1) = 0.$$

Thus this periodic orbit is superattracting. We have used the above algorithm to sketch the filled Julia set (and several magnifications) in Figure 2.

Note the fractal nature of the boundary of $K(P_{-1})$ which, of course, is $J(P_{-1})$. With the exception of $c = 0$ and $c = -2$, it is known that $J(P_c)$ is a fractal.

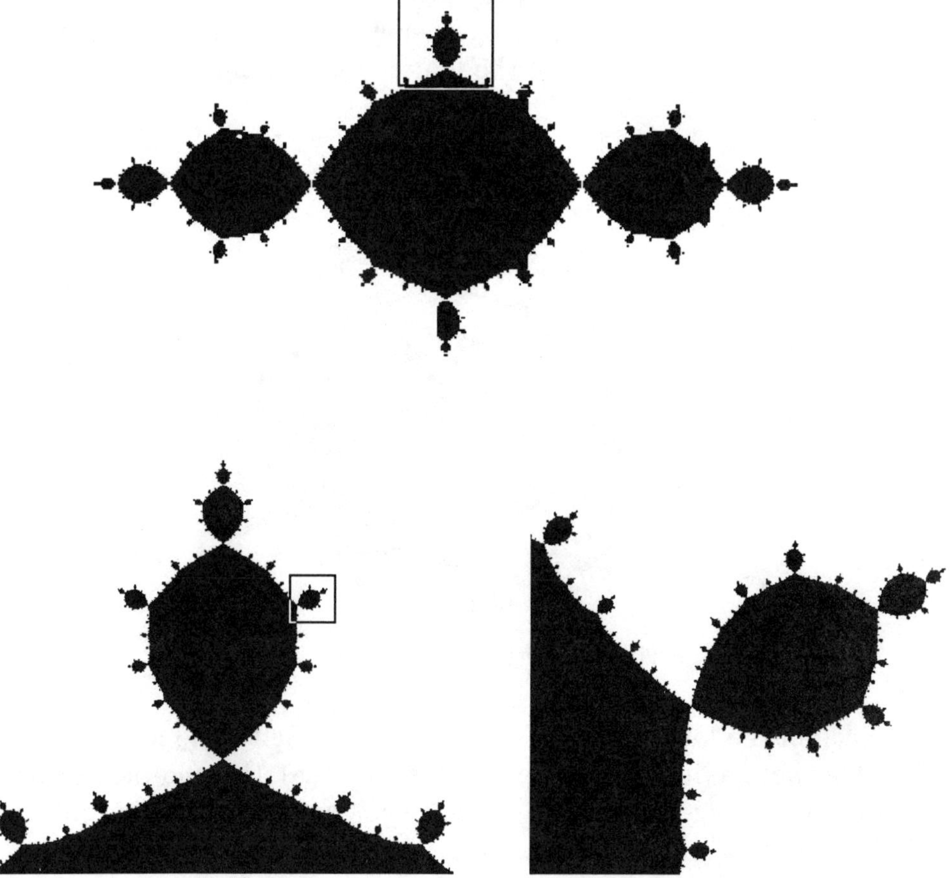

Fig. 2 The filled Julia set of P_{-1} and several magnifications.

Example 4. $P_5(z) = z^2 + 5$. Actually this example is quite typical. The following arguments hold for all c for $|c|$ sufficiently large (for example, $|c| > 2$ works).

Consider first the region $R_0 = \{z \mid |z| \leq 5\}$. Note that the critical value $P_5(0) = 5$ lies on the boundary of R_0. The image of R_0 is a disk of radius 25 centered at 5. Hence R_0 is contained in the interior of $P_5(R_0)$. Moreover, it is easy to check that the annular region $P_5(R_0) - R_0$ is a *fundamental domain* for the superattracting fixed point at ∞. This means that the orbit of each point in $P_5(R_0) - R_0$ tends to ∞, and moreover any orbit that tends to ∞ meets this set exactly once. Thus, to find $K(P_5)$, we will first look for those points whose orbits enter $P_5(R_0) - R_0$. The complement of this escape set is $K(P_5)$.

Let $R_1 = \{z \mid P_5(z) \in R_0\}$. Note that $0 \in R_1$ and that 0 is the only preimage of 5 contained in R_0. All other points in R_0 have exactly two preimages in R_1. One may check that, as a consequence, the region R_1 is a figure-eight together with its interior as shown in Figure 3. We call the lobes of the figure-eight I_0 and I_1.

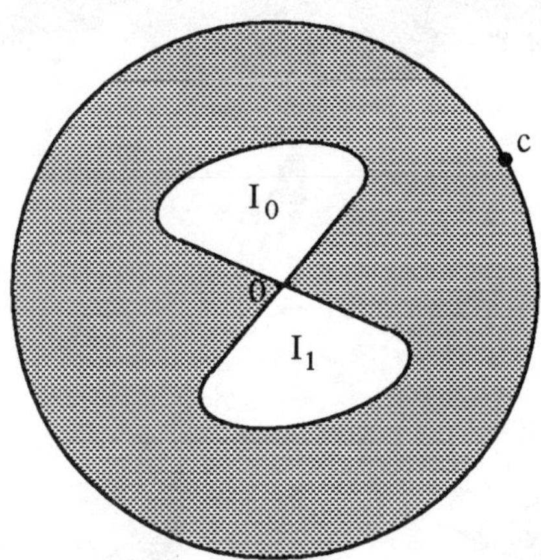

Fig. 3 The figure-eight region R_1 and its image.

Each of the lobes I_0 and I_1 is mapped in one-to-one fashion onto R_0. By the maximum principle, it follows that the complementary region $R_0 - R_1$ is mapped onto $P_5(R_0) - R_0$, and so each point outside R_1 has an orbit which escapes.

Now, since each lobe of R_1 is mapped onto R_0 in one-to-one fashion, it follows that there are a pair of subsets (one in each lobe) of R_1, each of which is mapped homeomorphically onto R_1. Call the union of these sets R_2. The set R_2 consists of a pair of figure-eights, I_{00} and I_{01} in I_0 and I_{10} and I_{11} in I_1. Then R_2 lies in the interior of R_1, and $P_5^2 : R_2 \to R_1$. Continuing inductively, let

$$R_{n+1} = P_5^{-1}(R_n).$$

It can be checked that R_{n+1} lies in the interior of R_n and that R_{n+1} consists of exactly 2^n components, each of which is homeomorphic to a figure-eight together with its interior. See Figure 4.

Finally, it follows that $K(P_5)$ is the nested intersection of the R_n, and so it follows that $K(P_5)$ consists of infinitely many components. In fact, it can be shown that each sequence of components of R_N nest down to a single point as $N \to \infty$ and that $K(P_5) = J(P_5)$ is a Cantor set.

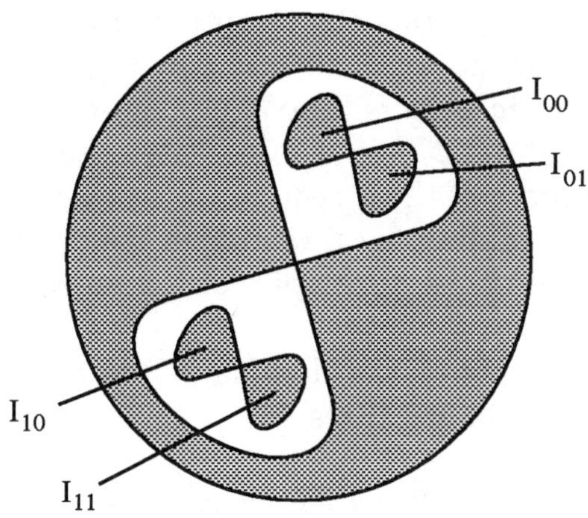

Fig. 4 $K(P_5)$ is a nested intersection of "figure-eights".

4. Role of the critical orbit. The orbit of the critical point plays a pivotal role in complex dynamics. There are several important examples of this. For example, recall that in the previous section we indicated that the Julia set of $P_5(z) = z^2 + 5$ was a Cantor set. There was nothing special about the value $c = 5$ in this argument. The essential ingredient in the proof is the fact that the orbit of 0 escapes. For any quadratic polynomial, similar arguments then give:

Theorem. *If the orbit of 0 escapes under iteration of P_c, then $K(P_c) = J(P_c)$ is a Cantor set.*

Actually, this theorem is the first part of an amazing dichotomy. If the orbit of 0 escapes for P_c, then $K(P_c)$ is totally disconnected, but if the orbit of 0 does not escape, then $K(P_c)$ is a connected set! That is, there are only two types of filled Julia sets for quadratic polynomials, those that consist of one component and those that consist of infinitely many (point) components. We will sketch a proof of this latter fact in the sequel, but for now we single it out as a theorem.

Theorem. *If the orbit of 0 does not escape under iteration of P_c, then both $K(P_c)$ and $J(P_c)$ are connected sets.*

This basic dichotomy allows us to define the Mandelbrot set.

Definition. The *Mandelbrot set* is given by $M = \{c \mid P_c^n(0) \not\to \infty\}$. Equivalently, the Mandelbrot set is the set of c-values for which $K(P_c)$ is a connected set.

This dichotomy highlights the important role played by the critical orbit in the dynamics of quadratic polynomials. The fate of the critical orbit defines in a gross topological sense the structure of the set of all bounded orbits. We will see later that the critical orbit determines much more of the structure of the filled Julia sets for P_c. Indeed, the critical orbits play a dominant role in any complex dynamical system. We will see examples of this in each chapter of this book.

As another example of the dominant role of the critical orbit, consider the following classical result.

Theorem. *If P_c has an attracting periodic orbit, then the orbit of 0 must be attracted to this orbit.*

Note the power of this theorem. Any quadratic polynomial must have infinitely many periodic points. However, at most one of them can be attracting!

We remark that this theorem is much more general. For example, if P_c has a rationally neutral cycle, then again the critical orbit must be attracted to it. We refer to the paper of Keen in this volume for more details about these facts.

As a final example, suppose the critical orbit is preperiodic. This means that the orbit of 0 lands on a cycle after some number of iterations, but 0 itself is not periodic. In this case there can be no attracting or neutral cycles for P_c, so in fact the Fatou set has no interior and we have $J(P_c) = K(P_c)$. Recall that this was the case in Example 2 of the previous section where, for P_{-2}, we saw that the orbit of 0 was eventually fixed:

$$0 \mapsto -2 \mapsto 2 \mapsto 2 \mapsto 2 \mapsto \cdots$$

Recall that the $J(P_c) = K(P_c)$ was the interval $[-2, 2]$ on the real axis.

As another example, consider $P_{-i}(z) = z^2 - i$. The orbit of 0 is preperiodic:

$$0 \mapsto -i \mapsto -i - 1 \mapsto i \mapsto -i - 1 \mapsto i \cdots$$

The Julia set for this map is a *dendrite* (an infinitely branched curve) as depicted in Figure 5.

Values of c for which the orbit of 0 is pre-periodic will be important when we discuss the geometry of the Mandelbrot set.

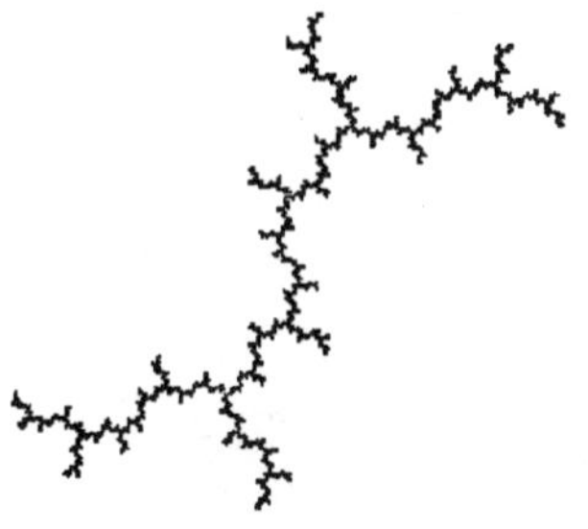

Fig. 5 The Julia set for P_{-i}.

Definition. A value of c for which the orbit of 0 under P_c is preperiodic (but not periodic) is called a *Misiurewicz point.*

5. External rays. The key to understanding the structure of both the Mandelbrot set and the Julia sets of quadratic polynomials is the theory of external rays developed by Douady and Hubbard in the early 1980s. It is known that any complex analytic map with a superattracting fixed point is locally conjugate in a neighborhood of this fixed point to $z \mapsto z^n$ for some $n \geq 2$. In our case, the point at ∞ is superattracting. Since P_c has degree 2, it follows that the map is conjugate of $z \to z^2$ near ∞. To be precise, for $R > 1$, let

$$V_R = \{z \in \mathbf{C} \,|\, |z| > R\}.$$

Then the result is

Theorem. *Let $c \in \mathbf{C}$. Then there exists a neighborhood U_c of ∞ in $\overline{\mathbf{C}}$, $R > 1$, and an analytic isomorphism $\phi_c \colon U_c \to V_R$ such that*

$$\phi_c(P_c(z)) = (\phi_c(z))^2.$$

Let us give a sketch of the proof. For $|z|$ sufficiently large, we will write down ϕ_c explicitly. For $|z| > \max(|c|, 2)$, define

$$\phi_c(z) = \lim_{n \to \infty} (P_c^n(z))^{1/2^n}.$$

The only difficulty with this definition is to specify which 2^n-th root to choose. But we may write

$$(P_c^n(z))^{1/2^n} = z \cdot \frac{(P_c(z))^{1/2}}{z} \cdot \frac{(P_c^2(z))^{1/4}}{(P_c(z))^{1/2}} \cdots \frac{(P_c^n(z))^{1/2^n}}{(P_c^{n-1}(z))^{1/2^{n-1}}}. \qquad (*)$$

In this expression, for each $k \leq n$, we may substitute

$$\frac{(P_c^k(z))^{1/2^k}}{(P_c^{k-1}(z))^{1/2^{k-1}}} = \left(1 + \frac{c}{(P_c^{k-1}(z))^2}\right)^{1/2^k}.$$

We may therefore choose the principal branch of this root provided each of $z, P_c(z), \ldots, P_c^k(z)$ are sufficiently large. Our computation after the escape time algorithm above shows that this will be the case when $|z| > \max(|c|, 2)$. Note that, using $(*)$, we clearly have

$$\phi_c(P_c(z)) = (\phi_c(z))^2,$$

so ϕ_c is a conjugacy. Moreover, it can be shown that ϕ_c is analytic in both z and c.

This defines ϕ_c in a neighborhood of ∞ in $\overline{\mathbf{C}}$. We may extend ϕ_c to a larger domain via the conjugacy. This is accomplished as follows. Suppose $0 \notin \partial U_c$. Let $z \in \partial U_c$. There exists $z' \in \partial U_c$ with $z \neq z'$ and $P_c(z) = P_c(z') \in U_c$. Choose a neighborhood W of $P_c(z)$ so that the preimage of W consists of two open sets W_1 and W_2 satisfying

   1. $z \in W_1$, $z' \in W_2$

   2. $W_1 \cap W_2 = \emptyset$.

The second condition is possible provided $P_c(0) = c \notin W$. Extend ϕ_c to all of W_1 and W_2 in a natural way by requiring $(\phi_c(z))^2 = \phi_c \circ P_c(z)$ and the continuity of ϕ_c. See Figure 6.

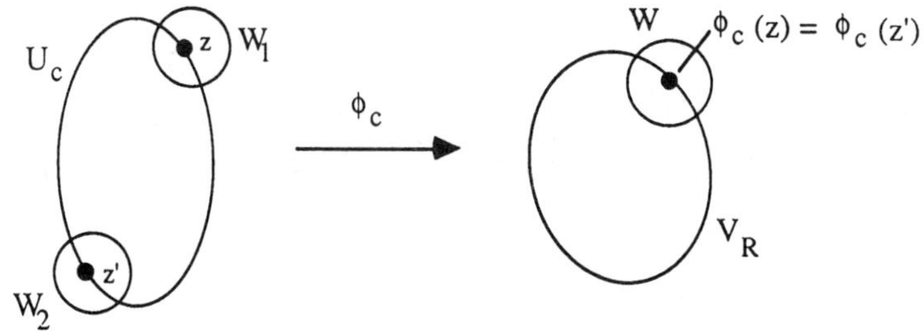

Fig. 6 Extending ϕ_c

Note that this procedure allows us to extend ϕ_c to an open neighborhood of ∞ that contains the critical point in its boundary if, in fact, the orbit of 0 escapes to ∞. If this is not the case, then the above procedure gives us an analytic isomorphism $\phi_c \colon \overline{\mathbf{C}} - K(P_c) \to \overline{\mathbf{C}} - \{z \mid |z| \leq 1\}$. That is, ϕ_c gives a uniformization of the complement of the filled Julia set of P_c in $\overline{\mathbf{C}}$. In particular, this shows that $K(P_c)$ is connected if $P_c^n(0) \not\to \infty$, a fact noted above.

If the orbit of 0 does escape, the above procedure allows us to define $\phi_c(c)$. Suppose $|\phi_c(c)| = r$. Then ϕ_c may be defined on an open neighborhood U of ∞ which contains 0 in its boundary such that $\phi_c \colon U \to V_{\sqrt{r}}$ is an isomorphism. Now $\overline{\mathbf{C}} - U$ is a figure-eight region, so this gives an analytic version of the topological picture we described in the previous section.

Of principal importance in the study of quadratic dynamics are external rays. These are the preimages of the straight rays in $\overline{\mathbf{C}} - V_{\sqrt{r}}$ under ϕ_c.

Definition. The *external ray* of argument θ is the curve $\gamma_\theta(t) = \phi_c^{-1}(te^{2\pi i\theta})$.

If $K(P_c)$ is connected, then each external ray accumulates on the boundary of $K(P_c)$, i.e., on the Julia set.

Example. Recall that the map $H(z) = z + 1/z$ gives a uniformization $H \colon \{z \mid |z| > 1\} \to \mathbf{C} - [-2, 2]$ that conjugates $z \mapsto z^2$ to $z \mapsto z^2 - 2$. This map determines the external rays, which are the pieces of hyperbolas in Figure 1.

If the Julia set is locally connected, then it is a fact that each external ray limits on a unique point in $J(P_c)$. We say that γ_θ *lands* at z if $\lim_{t\to 1} \gamma_\theta(t) = z$. Where the external rays land in $J(P_c)$ tells us a lot about the topology and the dynamics on the Julia set.

Unfortunately, not all Julia sets are locally connected, so it is not always the case that all external rays land. However, in a large number of cases it is known that all external rays do land on the Julia set. For example, if P_c has an attracting cycle, or if the orbit of 0 under P_c is eventually periodic, then it is known that $J(P_c)$ is locally connected.

6. The Mandelbrot set. Recall that the Mandelbrot set M is the set of all c-values for which the orbit of 0 under P_c is bounded. Which c-values lie in M? Certainly any c for which P_c has an attracting cycle lies in M, since, as we saw above, an attracting cycle always attracts the orbit of 0. Moreover, this situation is stable in the following sense: if P_c has an attracting cycle of period n, then there is a neighborhood U of c in the plane for which, if $c' \in U$, then $P_{c'}$ also has an attracting cycle of period n which is close to that of P_c. It follows that any such c-value lies in the interior of M.

Certain components of the interior of M consist of c-values for which P_c has an attracting cycle of some given period. These are called *hyperbolic* components. It is conjectured that *all* components of the interior of M are hyperbolic.

Example. The region C_1 where P_c has an attracting fixed point is the large cardioid-shaped region in M. See Figure 8. This region is determined by simultaneously solving the equations

$$P_c(z) = z^2 + c = z$$

$$|P_c'(z)| = |2z| < 1.$$

It is easy to check that this region is bounded by the curve

$$c = \zeta(\theta) = \frac{1}{2}e^{i\theta} - \frac{1}{4}e^{2i\theta}$$

where $0 \leq \theta < 2\pi$. For c-values on this curve, P_c has a neutral fixed point, and, if $c = \zeta(\theta)$, then the derivative of P_c at the fixed point is $e^{i\theta}$. Hence, the derivative wraps once around the unit circle as c traverses the boundary of the cardioid.

Example. The region C_2 where P_c has an attracting cycle of period 2 is bounded by the circle given by $|c + 1| = 1/4$. Again, it is easy to solve the equations

$$P_c^2(z) = z$$

$$|(P_c^2)'(z)| < 1$$

for c. Note that the boundaries of C_1 and C_2 meet at the unique point $c = -3/4$. As c passes from C_1 to C_2 through this point, the family of functions P_c undergoes a period-doubling bifurcation.

This scenario is typical. As the parameter c traverses the boundary of C_1, at each point where the derivative of P_c at the neutral fixed point is of the form $\exp(2\pi i\, p/q)$, we expect the family to undergo a period q bifurcation. Attached to C_1 at this point is a hyperbolic component in which P_c has an attracting cycle of period q. The point of attachment of two adjoining hyperbolic components is called a *root point*. We call the small bulb attached to C_1 at the p/q root point the p/q-*bulb*. Note that these bulbs can never overlap, since P_c has at most one attracting cycle.

In Plates 1-2, we have displayed the 2/5 and 10/21-bulbs. Note that each of the bulbs is decorated with a large "antenna." In each case, the antenna has a different topological structure. Indeed, each antenna features a central "junction point" from which various spokes emanate. If you count these spokes (including the main spoke emanating from the bulb, you will see that there are exactly q spokes in the p/q-bulb. This is the beginning of the fascinating story of the combinatorics of the Mandelbrot set which we will discuss below.

The numerator in the fraction p/q also plays a role in the dynamics of P_c for c in the p/q bulb. It is known that the filled Julia set for P_c contains a fixed point p at which exactly q components of $K(P_c) - \{p\}$ meet. The mapping P_c rotates these components about p by an angle $2\pi \cdot p/q$. Thus p/q may be thought of as a rotation number for this bulb2.

This fact is illustrated for the 1/4-bulb and the 2/5-bulb in Figure 7.

Fig. 7 The dynamics of P_c for c in the 1/4-bulb and the 2/5-bulb attached to C_1.

As is clear from Figure 8, there are a great many bulbs attached to each hyperbolic component in M. The following facts are known.

Theorem. *Suppose U is a component of the interior of M in which P_c has an attracting cycle of period K, say, $z_1(c), \ldots, z_K(c)$. Then*

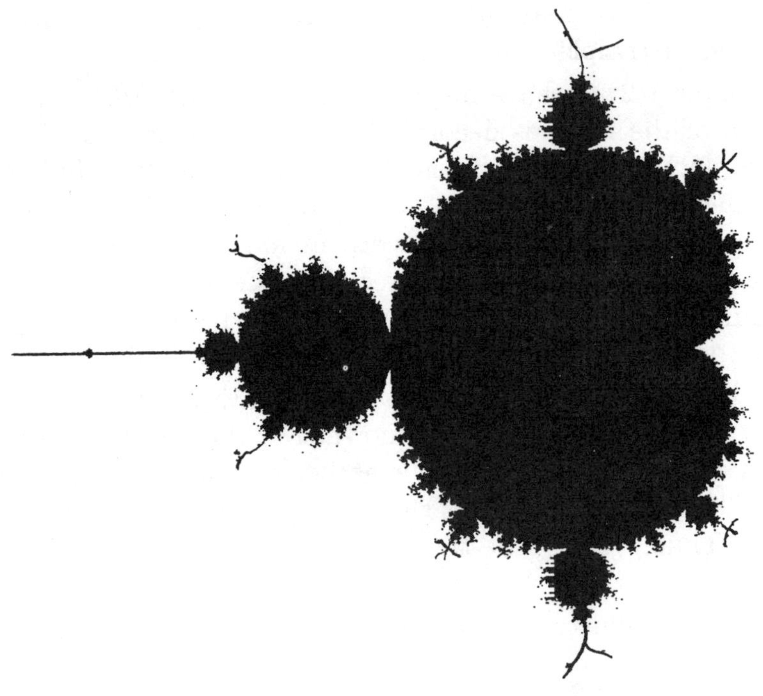

Fig. 8 The Mandelbrot Set.

1. The map $c \mapsto P_c'(z_i(c))$ is an analytic isomorphism of U onto the open unit disk for each i, $1 \leq i \leq k$.

2. As c winds once around the boundary of U, the derivative $P_c'(z_i(c))$ winds once around the unit circle. Here $z_i(c)$ is a point on a neutral cycle for P_c.

Thus the local structure of M around the boundary of a hyperbolic component is quite complicated but nevertheless analytically predictable. However, there are many other types of points in (the boundary of) M. For example, recall that c is a Misiurewicz point if the orbit of 0 is preperiodic under P_c. It is known that the set of Misiurewicz points is dense in the boundary of M. Similarly, accumulation points of hyperbolic components give the entire the boundary of M. More precisely, the boundary of M is contained in the closure of the set of centers of hyperbolic components. Here the *center* of a hyperbolic component is the unique c-value in that component for which the attracting cycle is superattracting. The proofs of each of these statements are easy consequences of Montel's Theorem. See [B].

Another intriguing aspect of the Mandelbrot set is the presence of small copies of M embedded within the larger set. Each of these small copies possesses a central cardioid in which P_c has an attracting cycle of some period, say k. Directly attached to this cardioid are infinitely many bulbs with period a multiple of k, just as in the case of the main cardioid. For example, there is a small Mandelbrot set of period 3 lying along the negative real axis with cusp at $c = -7/4$. In fact, there are infinitely many small copies of the Mandelbrot set in M. Plates 3-4 show small copies of M that lie in the 2/5 and 10/21-bulbs respectively. These are the largest copies of M visible in Plates 1-2. Note that, while the basic shape of these small Mandelbrot sets is the same in each case, the antenna structure differs markedly. Indeed, if you count the number of pieces of antenna emanating from junction points visible in these plates, you will see that there are exactly q such pieces, where q is the denominator of the fraction.

Thus the natural question is: How do we understand the global structure of M? The answer to this question is quite beautiful. Following Douady and Hubbard [DH1], we invoke the theory of external rays.

Recall that, for each $c \in \mathbf{C}$, there is defined a map ϕ_c which conjugates the dynamics of P_c near ∞ to $z \mapsto z^2$ outside of some large circle. If $c \notin M$, then $P_c^n(0) \to \infty$ as $n \to \infty$ and, as we saw in the previous section, ϕ_c may be extended in a natural way so that $\phi_c(c)$ is defined. Thus we may define another map $\Phi \colon \mathbf{C} - M \longrightarrow \mathbf{C} - \{z \mid |z| \le 1\}$. It is a remarkable fact (due to Douady and Hubbard) that this map gives a uniformization of the exterior of M.

Theorem. *The map* $\Phi(c) = \phi_c(c)$ *for* $c \in \mathbf{C} - M$ *is an analytic isomorphism onto* $\mathbf{C} - \{z \mid |z| \le 1\}$.

As a corollary, we obtain

Corollary. *The Mandelbrot set is a connected set.*

The map Φ allows us to define external rays in the complement of M.

Definition. *The external ray of M with angle θ is the curve* $R_\theta(t) = \Phi^{-1}(te^{2\pi i\theta})$.

As in the dynamical plane, we may again ask if all external rays of M land at a unique point on the boundary of M. This would be the case if M were locally connected. However, this is an open question. In fact, the main conjecture concerning M is that M is locally connected. While a proof of this fact is not yet known, there has been much recent progress by mathematicians such as Lyubich, McMullen, Shishikura, Swiatek, and Yoccoz.

                          R. L. DEVANEY

It is true that certain of the external rays do land on M. For example, all rational external rays are known to land. More importantly, the dynamics of P_c at the landing point is effectively determined by the argument of these rays.

Theorem. *Suppose $\theta = p/q$ in lowest terms. Then the external ray of M with angle θ lands at a point c_θ in the boundary of M. Moreover,*

    1. If q is odd, then c_θ is a root point of a hyperbolic component.
    2. If q is even, then c_θ is a Misiurewicz point.

In the first case, since q is odd, p/q has a repeating binary expansion, $B(p/q) = \overline{s_1 \ldots s_n}$ where $s_j = 0$ or 1 for each j. Then the root point at which the p/q ray lands separates a hyperbolic component of period n from a lower period component (with period dividing n), unless this root point is the cusp of a cardioid of one of the small copies of the Mandelbrot set in M.

Example. The binary expansions of $1/3$ and $2/3$ are $\overline{01}$ and $\overline{10}$, respectively. These rays must therefore land at the root point of the period 2 bulb, i.e., $c = -3/4$.

Example. In binary, we have

$$B(1/7) = \overline{001}, \ \ B(2/7) = \overline{010}, \ \ B(3/7) = \overline{011},$$

and so forth. So the external rays of angle $p/7$ must land at root points for period 3 components. Two of these are bulbs attached to the main cardioid where we have a period tripling bifurcation. The other must be a root point of a "stand-alone" period 3 component, i.e., a small copy of a Mandelbrot set within M which has period 3. This is because period 3 bulbs can only be attached to the main cardioid by direct bifurcation, not to bulbs with other periods. We have already seen that there is such a component in M along the negative real axis, hence the rays $3/7$ and $4/7$ must terminate here. See Figure 9.

Exercise. In binary, $p/15$ is given by a repeating sequence of period 4. Using the fact that distinct external rays cannot cross (except at their landing points) together with the rays already determined, you should be able to determine the approximate locations of the landing points for these rays. What can you say about the periods of the bulbs at these root points? What can you then say about external rays with argument $p/31$ (i.e., repeating sequences of length 5 in binary).

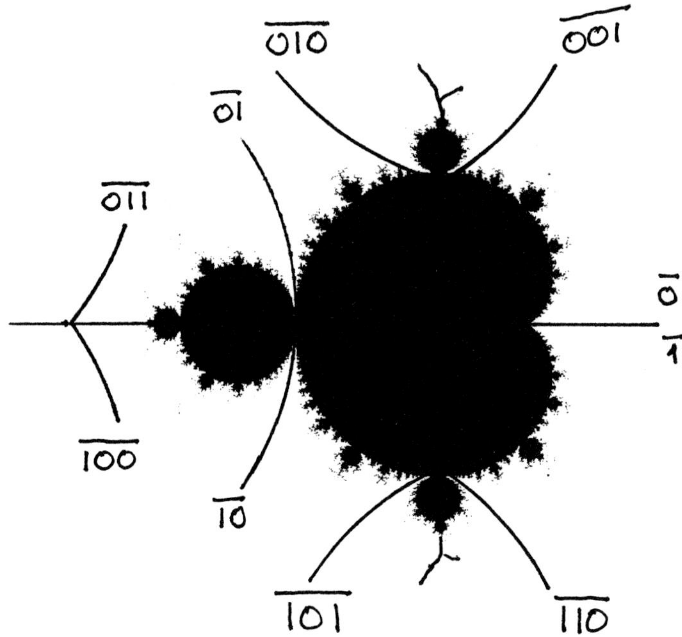

Fig. 9 External rays landing on the Mandelbrot set.

Schleicher's Algorithm. There is a neat algorithm to determine the rays which land at the p/q bulbs attached to the main cardioid due to D. Schleicher. To begin, we note that the ray given in binary by $\overline{0} = \overline{1}$ lands at the cusp of the main cardioid, i.e., at the root point of the 0/1 or 1/1 bulb. Also, the rays $\overline{01}$ and $\overline{10}$ land at the root of the 1/2 bulb.

Now we begin the inductive procedure. We first determine the rays that land at the "largest" bulb between the 0/1 and 1/2 bulb. By definition, the largest bulb is given by Farey addition, i.e.,

$$\frac{0}{1} \oplus \frac{1}{2} = \frac{1}{3}.$$

To determine the rays that land at the 1/3 bulb we note that the two closest rays already constructed are $\overline{0}$ (attached to 0/1) and $\overline{01}$ (attached to 1/2). To get the ray attached to the 1/3 bulb closest to the cusp, we first write the nearest ray constructed, 0, then the further ray, 01. This gives 001. Repeating, we find $\overline{001}$ is the ray attached to the 1/3 bulb closest to the cusp.

For the other ray, we first write the nearer ray already constructed, namely 01, then the further 0. Thus, $\overline{010}$ is the second ray attached to the 1/3 bulb's root point.

In general, Schleicher's algorithm allows us to find the two rays attached to the "largest" bulb between a p/q and r/s bulb, assuming the rays attached to these bulbs have already been constructed.

Step 1 Determine the largest bulb by Farey addition, i.e.,

$$\frac{p}{q} \oplus \frac{r}{s} = \frac{p+r}{q+s}.$$

Step 2 Find the rays closest to the $(p+r)/(q+s)$ bulb, say, $\overline{s_1 \ldots s_n}$ (attached to the p/q-bulb) and $\overline{t_1 \ldots t_m}$ (attached to r/s).
Step 3 The ray attached to the $(p+r)/(q+s)$ bulb closest to the p/q bulb is $\overline{s_1 \ldots s_n \, t_1 \ldots t_m}$. The ray attached to the $(p+r)/(q+s)$ bulb closest to the r/s bulb is $\overline{t_1 \ldots t_m \, s_1 \ldots s_n}$.

In this fashion, we can determine all rays attached to bulbs attached to the main cardioid. See Figure 10.

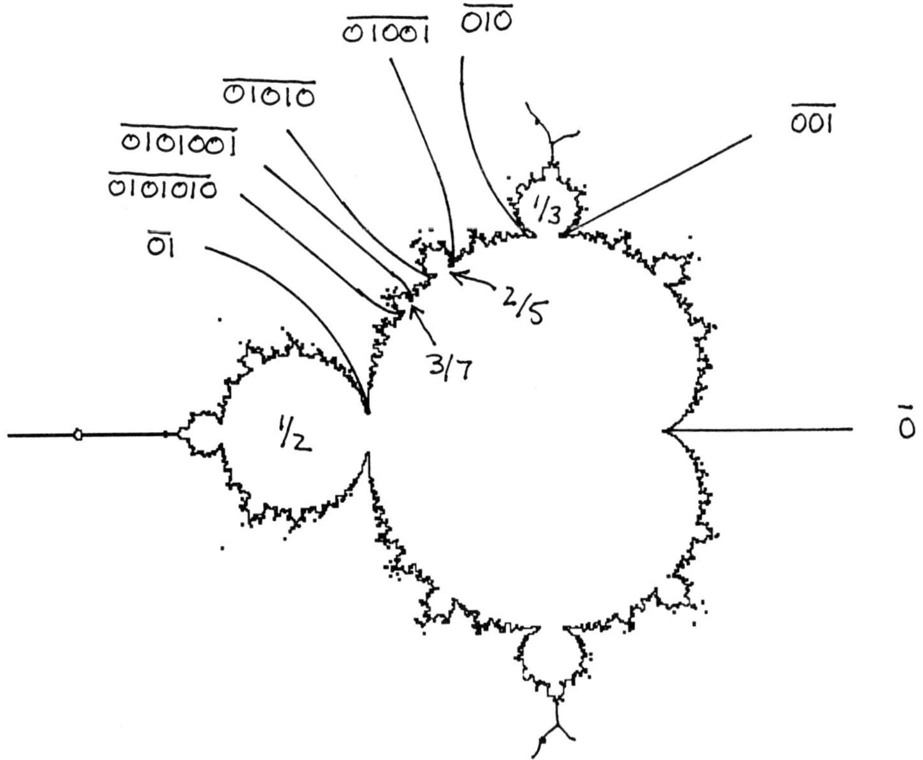

Fig. 10 Schleicher's algortithm.

Tuning. There is another algorithm that allows us to extend the Schleicher procedure to find the rays that land at many other root points. This algorithm is called *tuning*.

Suppose we have a hyperbolic component K with period k and know that the rays with binary expansions $\overline{s_1 \ldots s_n}$ and $\overline{t_1 \ldots t_n}$ land at its root point. Suppose also that, in binary, $\overline{s_1 \ldots s_n} < \overline{t_1 \ldots t_n}$. It is a fact that there is a homeomorphism $h\colon M \to M$ taking the main cardioid C_1 onto K and taking bulbs directly attached to C_1 onto those directly attached to K. The map h is by no means surjective, but it is important to note that K need not be a cardioid-shaped region. For example, there is a homeomorphism that takes M into M and maps C_1 onto any given p/q-bulb directly attached to C_1.

Now if we know where a particular ray lands at a point $c \in M$, we can compute the ray that lands at $h(c)$ by the following procedure. Suppose $\overline{\alpha_1 \ldots \alpha_j}$ lands at c. Then the ray landing at $h(c)$ with analogous orientation is obtained by replacing each α_i by the string $\overline{s_1 \ldots s_n}$ if $\alpha_i = 0$ or by $\overline{t_1 \ldots t_n}$ if $\alpha_i = 1$.

Example. Consider the $1/3$-bulb in M. The rays landing at its root point are $\overline{001}$ and $\overline{010}$, respectively. To find the rays that land at the period-doubling point attached to this bulb (i.e., the unique period 6 bulb directly attached to the $1/3$ bulb), we recall that the rays $\overline{01}$ and $\overline{10}$ land at the period-doubling point on C_1. Therefore, the two rays landing at the period-doubling point on the $1/3$-bulb are $\overline{001010}$ and $\overline{010001}$. In Figure 11 we display several other rays attached to the $1/3$-bulb as determined by tuning.

Since Schleicher's algorithm generates all of the rays attached to the main cardioid at root points, tuning allows us to find any ray that lands at a root point which is attached to C_1 by a finite sequence of bifurcations.

There are a number of other studies of how the rays land on the Mandelbrot set. In particular, the study of how the external rays in the dynamical plane change their destinations as c varies outside of the Mandelbrot set is fascinating. We refer to [At] for details.

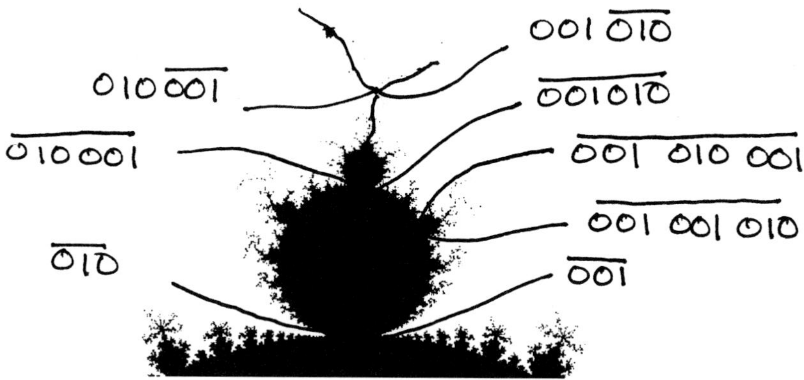

Fig. 11 Tuning the 1/3-bulb.

References

[Al] Alexander, D. Thesis Boston University, 1992.

[At] Atela, P. Bifurcations of Dynamic Rays in Complex Polynomials of Degree Two. *Ergod. Th. & Dynam. Sys.* **12** (1991), 401-423.

[Ba] Baker, I. N. Wandering Domains in the Iteration of Entire Functions. *Proc. London. Math. Soc.* **49** (1984), 563-576.

[B] Blanchard, P. Complex Analytic Dynamics on the Riemann Sphere, *B.A.M.S.* Vol. II, No.1, 1984, 85-141.

[BBCDS] Banks, J., Brooks, J., Cairns, G., Davis, G., and Stacey, P. On Devaney's Definition of Chaos. *Amer. Math. Monthly* **99** (1992), 332-334.

[Bra] Branner, B. The Mandelbrot Set. In *Chaos and Fractals: The Mathematics Behind the Computer Graphics.* Amer. Math. Soc. (1989), 75-106.

[Bro] Brolin, H., Invariant Sets under Iteration of Rational Functions, *Ark. Math.* **6** (1965), 103-144.

[D] Devaney, R. L. *An Introduction to Chaotic Dynamical Systems*, Second Edition. Addison-Wesley Co., Redwood City, Calif., 1989.

[D1] Devaney, R. L. *A First Course in Chaotic Dynamical Systems*, Addison-Wesley, Co., Reading, MA, 1992.

[DH] Douady, A. and Hubbard, J. Étude Dynamique des Polynôme Complexes, *Publications Mathematiques d'Orsay*, 84-102.

[DH1] Douady, A. and Hubbard, J. Itération des Polynômes quadratiques complexes, *C.R. Acad. Sci. Paris*, t.29, Serie I-1982, pp. 123-126.

[F] Fatou, P., Sur l'Itération des fonctions transcendentes Entières, *Acta Math.* **47** (1926), 337-370.

[K] Keen, L. Julia Sets. In *Chaos and Fractals: The Mathematics Behind the Computer Graphics.* Amer. Math. Soc. (1989), 57-74.

[S] Sullivan, D., Quasiconformal Maps and Dynamical Systems I, Solutions of the Fatou-Julia Problem on Wandering Domains. *Ann. Math.* **122** (1985), 401-418.

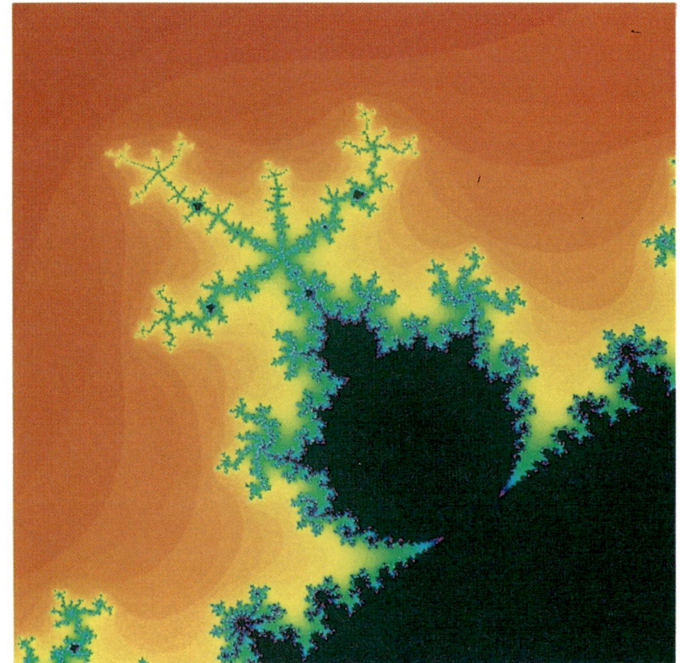

COLOR PLATE 1. The 2/5 bulb.

COLOR PLATE 2. The 10/21 bulb.

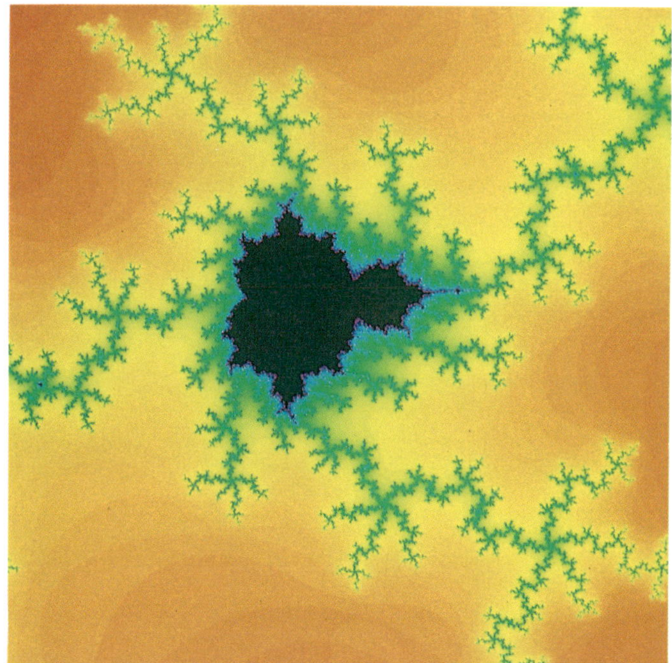

COLOR PLATE 3. Small copy of M attached to the 2/5 bulb.

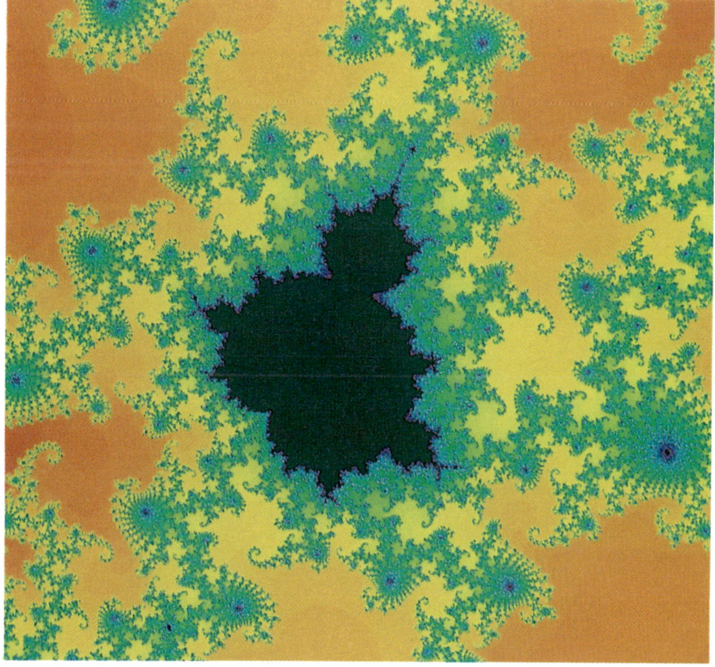

COLOR PLATE 4. Small copy of M attached to the 10/21 bulb.

Proceedings of Symposia in Applied Mathematics
Volume **49**, 1994

Puzzles and Para-Puzzles
of Quadratic and Cubic Polynomials

Bodil Branner

ABSTRACT. The paper reviews the construction of puzzles of quadratic and cubic polynomials with one bounded critical orbit. The similarities in the constructions are emphasized. In the quadratic case the goal is to prove local connectivity of the Julia set; in the cubic case the goal is to characterize the polynomials with totally disconnected Julia set. The constructions of para-puzzles are also briefly reviewed.

1. Introduction

Many global dynamical properties can be derived from a knowledge of the behavior of the critical points under iteration. For instance, the Julia set of a polynomial is connected if and only if all critical points have bounded orbits and totally disconnected if all critical orbits are unbounded. Since a quadratic polynomial has only one critical point, the Julia set is either connected or totally disconnected. For a polynomial of degree $d > 2$ there is another possibility: some critical orbits may be bounded and others unbounded. The simplest such case is achieved by cubic polynomials with one bounded critical orbit and one unbounded critical orbit. The Julia set must be disconnected; in some cases it is totally disconnected, in other cases not.

The puzzles and the tableaux technique was developed in [**BH2**] in order to characterize the cubic polynomials with one bounded and one unbounded critical orbit having totally disconnected Julia set. This technique has also been applied by Yoccoz (though formulated in another language) to certain quadratic polynomials with connected Julia set in order to prove local connectivity of the Julia set ([**Y**], [**H**], [**M**]). In both cases the global property (i.e. that the Julia set is totally disconnected or that the Julia set is locally connected) is derived from the dynamical behavior of the critical point with bounded orbit. We shall describe the construction of the puzzles and of the tableaux technique in both cases. The cubic case is less technical than the quadratic case. We shall emphasize the similarities and point out where the differences occur.

1991 Mathematics Subject Classification. Primary 58F23: Secondary 30D05.

The results in the dynamical plane are transferred to results in corresponding one-dimensional parameter spaces. In the quadratic case Yoccoz has proved local connectivity of the Mandelbrot set at the points corresponding to the quadratic polynomials with locally connected Julia set mentioned above. In the cubic case the analogous result in the appropriate one-dimensional parameter space is that the point corresponding to a cubic polynomial with totally disconnected Julia set as mentioned above is a point component in the set of cubic polynomials with one bounded and one unbounded critical orbit. We shall not go into many details in the parameter spaces, but concentrate on the description of the successive subdivision of the parameter spaces: the construction of the para-puzzles. The subdivision in the quadratic case is the easiest to follow, since the total parameter space is one-dimensional, but again the quadratic case is more technical than the cubic case.

The common strategy for proving that certain connected components are point components can be described as follows:
We start with a complex one-dimensional space which arises either as a dynamical plane ($\mathbb{C}$) or as a one-dimensional parameter space ($\mathbb{C}$ or a one-dimensional complex manifold). In this one-dimensional space we consider a compact subset K and a disconnected set $X \subset K$ (which may or may not equal K). For each $x \in X$ let $X(x)$ denote the connected component of X containing x. For $n = 0, 1, 2, \ldots$ we construct a closed neighborhood $P_n(x)$ of $X(x)$ homeomorphic to a closed disk and such that the set of neighborhoods form a nested sequence

$$P_0(x) \supset P_1(x) \supset \cdots \supset P_n(x) \supset \ldots$$

Furthermore we construct for $n = 0, 1, 2, \ldots$ an open annulus

$$A_n(x) \subset P_n(x) \setminus P_{n+1}(x).$$

When the strategy is successful we refer to it as *the method of divergence* meaning that the series formed by the moduli of these annuli is divergent

$$\sum_{n=0}^{\infty} \operatorname{mod} A_n(x) = \infty.$$

The important consequence derived from the divergence of this series is that the diameter of $P_n(x)$ tends to zero as n tends to infinity, hence that the connected component $X(x)$ equals one point, i.e. $X(x) = \{x\}$.
When the goal is to prove local connectivity of K at a point x then the neighborhoods $P_n(x)$ should be constructed so that $P_n(x) \cap K$ is connected. From this and the divergence we then derive that the sets $P_n(x) \cap K$ form a basis of connected neighborhoods of x; hence that K is locally connected at x.

When we work in the dynamical plane a neighborhood $P_n(x)$ is called a *puzzle-piece at depth n* and in the parameter space it is called a *para-puzzle-piece at depth n*.

This paper review the construction of puzzles and para-puzzles and the tableaux technique. The theorems about local connectivity in the quadratic case and the theorems about point components in the cubic case are stated after the description of puzzles and para-puzzles.

The parts about quadratic polynomials can be regarded as a continuation of the paper *The Mandelbrot Set* [**B**].

We will give very few proofs, but state the idea of a proof when it is possible within the framework. The reader is not assumed to be able to fill all the gaps, but is encouraged to dig into the references.

The paper consists of the following sections:
    2. The dynamical plane.
    3. Parameter spaces.
    4. The method of divergence.
    5. Puzzles.
    6. Para-puzzles.
    7. The main theorems.
    8. The tableaux technique.
    9. Final Remarks.

I want to thank Adrien Douady and John H. Hubbard for inspiration, explanations and discussions over the years, Dan Sørensen for providing several computer programs which were also used to produce computer pictures in this paper and Robin Theander for providing the Colored Plates 1 – 4 and the Figures related to those.

2. The dynamical plane

In this section we review some of the basic dynamical properties we shall use in the following.

2.1. Parametrization.

Any complex polynomial of degree $d \geq 2$ is conjugate to one of the form

$$f_{\mathbf{c}}(z) = z^d + c_{d-2}z^{d-2} + \cdots + c_1 z + c_0$$

by an affine change of the dynamical variable z. In particular we obtain quadratic polynomials of the form $f_c(z) = z^2 + c,\ c \in \mathbb{C}$, and cubic polynomials of the form $f_{(c_1,c_0)}(z) = z^3 + c_1 z + c_0,\ (c_1, c_0) \in \mathbb{C}^2$.
The *critical set*

$$\Omega(f_{\mathbf{c}}) \ = \ \{\ \omega \in \mathbb{C} \ | f'_{\mathbf{c}}(\omega) = 0\}$$

plays an important role. A polynomial of the form $f_{\mathbf{c}}$ is called *monic* and *centered*; monic because the leading coefficient is 1 and centered because

$$\sum_{\omega \in \Omega(f_{\mathbf{c}})} \omega = 0.$$

Different parameterizations are useful for different purposes. For our purpose it is important that the polynomials are monic. For convenience we shall most often write f instead of $f_{\mathbf{c}}$.

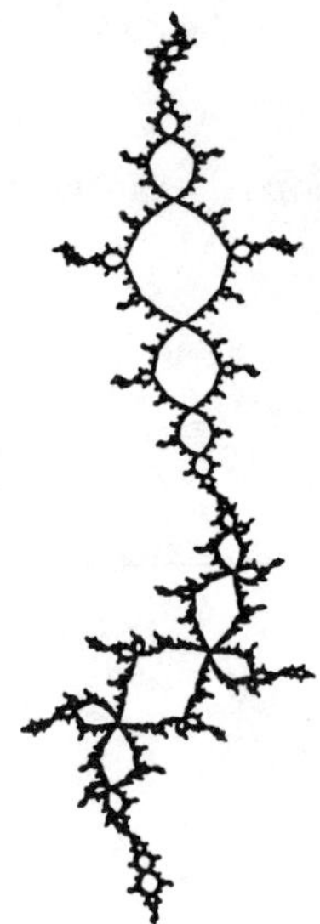

Figure 1. The Julia set of a cubic polynomial with bounded critical orbits. The critical points are marked.

The sequence

$$z_0, z_1 = f(z_0), \ldots, z_n = f(z_{n-1}), \ldots$$

is called the *orbit* of z_0 under iteration, and a sequence starting from a critical point is called a *critical orbit*.

2.2. The filled Julia set and the Julia set.

For each polynomial f of degree $d \geq 2$ the dynamical plane decomposes into two complementary sets: the set of points with bounded orbit and the set of points with unbounded orbit. We denote by K_f the set of seeds with bounded orbit, called the *filled Julia set*,

$$K_f = \{z \in \mathbb{C} \mid f^{\circ n}(z) \nrightarrow \infty\} \ .$$

We denote by $A_f(\infty)$ the set of seeds with unbounded orbit:

$$A_f(\infty) = \mathbb{C} \setminus K_f = \{z \in \mathbb{C} \mid f^{\circ n}(z) \to \infty\} \ .$$

The set $A_f(\infty)$ is called the *attractive basin* of ∞. The set $A_f(\infty)$ is completely invariant under iteration, i.e. under both forward and backward iteration. The common boundary

$$\partial K_f = \partial A_f(\infty) = J_f$$

is called the *Julia set*. The filled Julia set and the Julia set are compact subsets of $\mathbb{C}$. Note that a critical orbit is bounded if the critical point belongs to K_f and

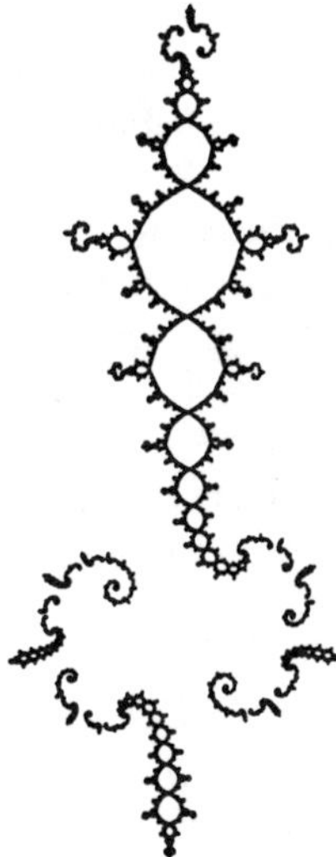

Figure 2. The Julia set of a cubic polynomial with one bounded
and one unbounded critical orbit. The critical points are marked.

unbounded if the critical point belongs to $A_f(\infty)$. Figures 1 - 3 give examples of
Julia sets of cubic polynomials.

The dynamical plane is also decomposed into two other complementary subsets: the
Fatou set on which the dynamics is *stable* and the Julia set on which the dynamics
is chaotic. A connected component of $\mathbb{C} \setminus J_f$ is called a *Fatou component*. We refer
to Keen [**K**] for a general introduction to Julia sets. Here we shall just mention

Figure 3. The Julia set of a cubic polynomial with unbounded
critical orbits. The critical points are marked.

that the Julia set is always a compact, perfect set in $\mathbb{C}$.

Remarks. (1) We say that K_f is *full* since it is bounded with connected complement $A_f(\infty)$.
(2) The filled Julia set K_f is obtained from the Julia set J_f by *filling in* all the bounded components of $\mathbb{C} \setminus J_f$, hence the name.
(3) It follows that the filled Julia set K_f is connected if and only if the Julia set J_f is connected.

The dynamics is dominated by the behavior of the critical points in $\mathbb{C}$. One classical theorem which supports this statement is the following:

THEOREM 1 (P. FATOU AND G. JULIA). *Let f denote a polynomial and $\Omega(f)$ its critical set. Then*

$$(1) \quad \Omega(f) \subset K_f \iff J_f \text{ is connected;}$$

$$(2) \quad \Omega(f) \subset A_f(\infty) \implies J_f \text{ is a Cantor set.}$$

Remarks. (1) A *Cantor set* is defined as a compact, perfect, totally disconnected subset in $\mathbb{C}$. Any such set is homeomorphic to the middle-third Cantor set and therefore deserves the name a Cantor set.
(2) Since any Julia set is compact and perfect it follows that a given Julia set is a Cantor set if and only if it is totally disconnected.

For a quadratic polynomial there is only one critical point. The Julia set (and the filled Julia set) is therefore either connected or a Cantor set. Compare also with Figures 1 and 3, where the Julia set is connected respectively totally disconnected. In Figure 2 the Julia set is disconnected, but not totally disconnected.

We shall comment on the proof of the first part of the theorem after we have introduced the Green function g_f of K_f and the Böttcher coordinates in a neighborhood of infinity and the second part of the theorem after we have introduced the method of divergence.

2.3. The Green function of the filled Julia set.
We define the function

$$g_f : \mathbb{C} \longrightarrow \mathbb{R}_+ \cup \{0\}$$

by

$$g_f(z) = \lim_{n \to \infty} \frac{1}{d^n} \log_+(|\, f^{\circ n}(z)\,|) \,,$$

where $\log_+(|\,z\,|) = \max\{0, \log(|\,z\,|)\}$. The mapping g_f is

(1) continuous on $\mathbb{C}$, harmonic on $\mathbb{C} \setminus K_f$ and equal to 0 on K_f .
(2) $g_f(z) = \log(|\,z\,|) + O(1)$ when $|\,z\,| \to \infty$,
(3) $g_f(f(z)) = d \cdot g_f(z)$.

Properties (1)-(2) show that the restriction of g_f to $\mathbb{C} \setminus K_f$ is the Green function of the filled Julia set K_f . The mapping g_f measures the *escape rate* to infinity. Note that not every compact set K in $\mathbb{C}$ gives rise to a Green function, but a compact set which arises as a filled Julia set of a polynomial does.

The maximal escape rate of the critical points plays an essential role; we set

$$G(f) = \max_{\omega \in \Omega(f)} \{g_f(\omega)\} .$$

2.4. The Böttcher coordinate in a neighborhood of ∞.

The point at ∞ is a super-attracting fixed point for any polynomial f extended to the Riemann sphere $\widehat{\mathbb{C}} = \mathbb{C} \cup \{\infty\}$. Set

$$U_f = \{z \in \mathbb{C} \mid g_f(z) > G(f)\}.$$

For a monic polynomial of degree d there exists a unique analytic isomorphism

$$\varphi_f : U_f \longrightarrow \mathbb{C} \setminus \bar{D}_{\exp G(f)}$$

satisfying

$$\frac{\varphi_f(z)}{z} \longrightarrow 1 \quad \text{as} \quad \mid z \mid \rightarrow \infty$$

and conjugating f to the simplest polynomial $f_0(z) = z^d$ of degree d , i.e.

$$\varphi_f \circ f = f_0 \circ \varphi_f .$$

Remarks. (1) Note that the two mappings g_f and φ_f are related by

$$g_f(z) = \log(\mid \varphi_f(z) \mid) \quad \text{for} \quad z \in U_f .$$

(2) If $\Omega(f) \subset K_f$ then $U_f = \mathbb{C} \setminus K_f$ and $\varphi_f : \widehat{\mathbb{C}} \setminus K_f \longrightarrow \widehat{\mathbb{C}} \setminus \overline{\mathbb{D}}$ defines a Riemann mapping. It follows that K_f is connected. This proves part of Theorem 1.

2.5. Equipotentials.

Choose any $\eta > 0$. The set $g_f^{-1}(\eta)$ is called the *equipotential* of value η.

A critical point of g_f is a point at which the gradient vector is zero. From $g_f(z) = \frac{1}{d} g_f(f(z))$ we obtain

$$\operatorname{grad} g_f(z) = \frac{1}{d} \operatorname{grad} g_f(f(z)) \cdot \overline{f'(z)}.$$

Since φ_f is a holomorphic isomorphism the Green function g_f of K_f has no critical points in U_f and any equipotential of value $\eta > G(f)$ is a simple closed curve surrounding K_f . A point z is a critical point of g_f if and only if

$$z \in \bigcup_n \bigcup_{\omega \in \Omega(f) \cap A_f(\infty)} f^{-n}(\omega) .$$

Hence if all the critical points of f are contained in K_f then g_f has no critical points in $A_f(\infty)$ and all equipotentials are simple closed curves surrounding K_f.

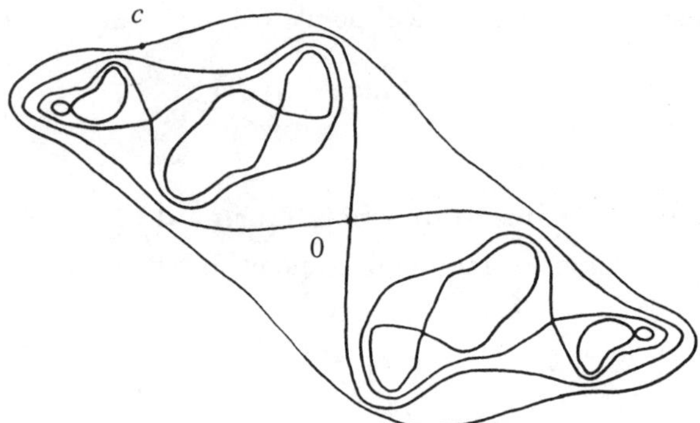

Figure 4. Some critical graphs of a quadratic polynomial with one unbounded critical orbit, together with the equipotential containing the critical value.

2.5.1. Critical graphs in the dynamical plane. Suppose f is a polynomial with at least one critical point escaping to infinity. An equipotential containing a critical point of g_f is called a *critical graph*. (Compare with Figure 4 and Figure 5.) The critical graphs are essential in the construction of the puzzles in the disconnected case, both for quadratic and cubic polynomials.

Remark. The critical graphs can also be defined without any reference to the potential g_f. Let ω be a critical point with unbounded orbit. For any $n \geq 0$ the

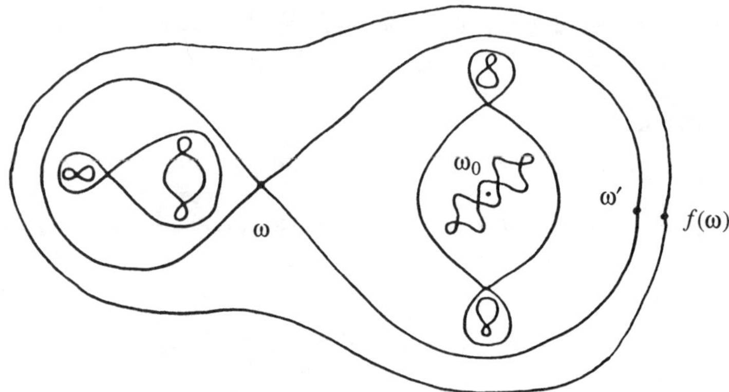

Figure 5. Some critical graphs of a cubic polynomial, together with the equipotential containing the critical value of the fastest escaping critical point.

set of points

$$\bigcup_{k=0}^{\infty} f^{-(k+n)}\left(f^{k}(\omega)\right)$$

form a dense set on the critical graph containing $z \in f^{-n}(\omega)$. Note that since the critical points are preserved under topological conjugation then the critical graphs are also preserved. It follows that the order in which critical points escape to infinity is unchanged under topological conjugation.

2.6. External rays.

Suppose K_f is connected. Let $\varphi_f : \mathbb{C} \setminus K_f \to \mathbb{C} \setminus \overline{\mathbb{D}}$ be as above. Set $\psi_f = \varphi_f^{-1}$. The orthogonal trajectories of the equipotentials are called *external rays*. The ray of *external argument* $\theta \in \mathbb{T} = \mathbb{R}/\mathbb{Z}$ is defined by

$$\mathcal{R}_f(\theta) = \psi_f(\{e^{\eta+2\pi i\theta}\}_{\eta>0}).$$

Note that the external ray of argument θ is mapped onto the external ray of argument $d \cdot \theta$:

$$f(\mathcal{R}_f(\theta)) = \mathcal{R}_f(d \cdot \theta).$$

An external ray is said *to land* if $\psi_f(\{e^{\eta+2\pi i\theta}\})$ has a limit as η tends to 0.

THEOREM 2 (DOUADY'S LANDING THEOREM). *Suppose K_f is connected. Each external ray of rational external argument $\frac{p}{q}$ do land. The landing-point is denoted* $\gamma_f(\frac{p}{q})$.

Some particular rational rays are essential in the construction of the puzzles in the connected quadratic case.

If K_f is not connected one can still consider the orthogonal trajectories to the equipotentials in $\mathbb{C} \setminus K_f$ minus the critical points of g_f . Some rays branch at critical points of g_f , the others continue unbroken towards K_f .

2.7. Local connectivity.

The concept of local connectivity and its implication – as stated below in the Carathéodory Theorem – are purely topological.

DEFINITION 3 (LOCAL CONNECTIVITY). *A set $X \subset \mathbb{C}$ is said to be locally connected at a point $x \in X$ if there exist arbitrary small connected neighborhoods of x in X, that is for each neighborhood U of x there exists a neighborhood $V \subset U$ of x such that $V \cap X$ is connected. The set X is said to be locally connected if it is locally connected at each point.*

THEOREM 4 (CARATHÉODORY). *Let $K \subset \mathbb{C}$ be compact, connected and full. Let $\varphi : \widehat{\mathbb{C}} \setminus K \to \widehat{\mathbb{C}} \setminus \overline{\mathbb{D}}$ be a Riemann mapping fixing ∞. Then $\psi = \varphi^{-1}$ can be extended to a continuous map $\varphi : \widehat{\mathbb{C}} \setminus \mathbb{D} \to \widehat{\mathbb{C}} \setminus \overset{\circ}{K}$ if and only if the set K is locally connected.*

If K is locally connected then the map $\gamma_K : \mathbb{T} \to \partial K$ determined by $\gamma_K(\theta) = \psi(e^{2\pi i\theta})$ gives a parametrization of the boundary of K, called the *Carathéodory loop*. Moreover, in this case one can construct a topological model of $K \subset \mathbb{C}$ from the data given by γ_K (see [**D**]). In that sense the compact, connected, locally connected, full subsets of $\mathbb{C}$ are well understood.

If in particular K equals the filled Julia set K_f of a polynomial f and K_f is locally connected then all external rays land and $\gamma_f(\theta) = \gamma_{K_f}(\theta)$ is the landing point of the ray $\mathcal{R}_f(\theta)$. The Theorem 2 above states that all rational rays land even if K_f is not locally connected. The symbol $\gamma_f(\frac{p}{q})$ of the landing point is now justified.

Remarks. (1) A non-connected Julia set cannot be locally connected since an arbitrary neighborhood of any point contains infinitely many components of the Julia set.

(2) The Julia set of a polynomial is known to be locally connected if it is connected and any of the critical points either has a finite orbit or is being attracted to an attracting periodic orbit or to a parabolic orbit. For quadratic polynomials there are further results. First of all the result by Yoccoz which is discussed in this paper. Yoccoz result gives a large class of locally connected Julia sets with a dynamics quite different from the ones mentioned above. Moreover, it has recently been proved in [**HJ**], [**J**] and [**L**] that the Julia set of some infinitely often renormalizable quadratic polynomials is locally connected and in [**P**] that the Julia set of a quadratic polynomial with a Siegel disk for which the rotation number is a Diophantine number with exponent 2 is locally connected.

3. Parameter spaces

In the following we introduce the one-dimensional parameter spaces for quadratic and cubic polynomials relevant for this paper.

3.1. The parameter space of quadratic polynomials.

For monic, centered quadratic polynomials $f_c(z) = z^2 + c$ the parameter space is the complex plane. We use the notion J_c, φ_c etc. instead of J_{f_c}, φ_{f_c} etc.

The *Mandelbrot set* $M \subset \mathbb{C}$ is the set of c-values for which the Julia set J_c is connected or equivalently for which the critical orbit is bounded.

THEOREM 5 (A.DOUADY AND J.H.HUBBARD). *(1) The map* $\Phi_M : \mathbb{C}\backslash M \to \mathbb{C}\backslash\overline{\mathbb{D}}$ *defined by* $\Phi_M(c) = \varphi_c(c)$ *is a conformal equivalence.*
(2) The Mandelbrot set is compact, connected and full.

The *hyperbolic* part HM of the Mandelbrot set consists of c-values for which the critical point is attracted to an attracting periodic orbit. The set HM is contained in the interior $\overset{\circ}{M}$. A connected component of HM is called a *hyperbolic component*. Let H denote a hyperbolic component. The multiplier map $\rho_H : H \to \mathbb{D}$ is defined by setting $\rho_H(c)$ equal to the multiplier of the unique attracting orbit of f_c. The map ρ_H is a conformal equivalence, and it extends to a homeomorphism $\rho_H : \overline{H} \to \overline{\mathbb{D}}$. The point $r_H = \rho_H^{-1}(1)$ is called the *root point* of H. The main hyperbolic component H_0 is the one bounded by the big cardioid.

There are two conjectures about the Mandelbrot set:

CONJECTURES 6. *(1) The Mandelbrot set is locally connected.*

(2) The hyperbolic part HM of the Mandelbrot set equals the interior $\overset{\circ}{M}$.

One can prove that if the first conjecture holds then the second does too. Moreover if M is locally connected then it follows from the Carathéodory Theorem that $\Psi_M = \Phi_M^{-1} : \mathbb{C} \setminus \overline{\mathbb{D}} \to \mathbb{C} \setminus M$ extends continuously to the boundary, giving rise to a Carathéodory loop γ_M.

Remark. The Mandelbrot set M is known to be locally connected at c-values in ∂M of the following types: boundary points of a hyperbolic component, c-values for which the critical point is strictly preperiodic under f_c ([**T**]) and finally c-values of the type discussed in this paper ([**Y**], [**H**]). The remaining set of points where M is not known to be locally connected corresponds to polynomials which are infinitely often renormalizable.

3.1.1. Equipotentials and external rays. We define the function

$$G_M : \mathbb{C} \to \mathbb{R}_+ \cup \{0\}$$

by $G_M(c) = g_c(c)$. Then $G_M(c) = \log |\Phi_M(c)|$ for $c \notin M$ and $G_M : \mathbb{C} \setminus M \to \mathbb{R}_+$ is the Green function of M.

As in the dynamical plane one defines an *equipotential* of value $\eta > 0$ to be the set $G_M^{-1}(\eta)$. Every equipotential is a simple closed curve.

An *external ray $\mathcal{R}_M(\theta)$ of external argument* $\theta \in \mathbb{T}$ is defined as

$$\mathcal{R}_M(\theta) = \Psi_M(\{e^{\eta + 2\pi i \theta}\}_{\eta > 0}).$$

Without knowing that M is locally connected one can still prove the landing of special external rays. In particular Douady and Hubbard have proved that all rational rays land. We shall need the result stated below about landing of rational rays, together with the result which relates landing properties in the dynamical plane to landing properties in the parameter space.

THEOREM 7 (A.DOUADY AND J.H.HUBBARD'S LANDING THEOREM). *(1) An external ray of rational argument of the form $\frac{p}{q}$ with $(p,q) = 1$ and q even lands and the landing point $c = \gamma_M(\frac{p}{q})$ corresponds to a polynomial for which the critical point is strictly preperiodic.*
(2) Moreover if c corresponds to a polynomial for which the critical point is strictly preperiodic then rays which land at c are rational of this form and $c = \gamma_M(\frac{p}{q})$ if and only if $c = \gamma_c(\frac{p}{q})$.
(3) An external ray of rational argument of the form $\frac{p}{q}$ with $(p,q) = 1$ and q odd lands and the landing point $c = \gamma_M(\frac{p}{q})$ is the root point of a hyperbolic component. If the hyperbolic component $H \neq H_0$ then there are exactly two rays which land at the root point r_H and they are rational; for H_0 there is exactly one ray which lands at the root point $r_{H_0} = \frac{1}{4}$, namely the ray $\mathcal{R}_M(0)$.
(4) Moreover if c belongs to the hyperbolic component H then $r_H = \gamma_M(\frac{p}{q})$ if and only if the external ray $\mathcal{R}_c(\frac{p}{q})$ is adjacent to the Fatou component containing c and the landing point $\gamma_c(\frac{p}{q})$ is the root point of this component.

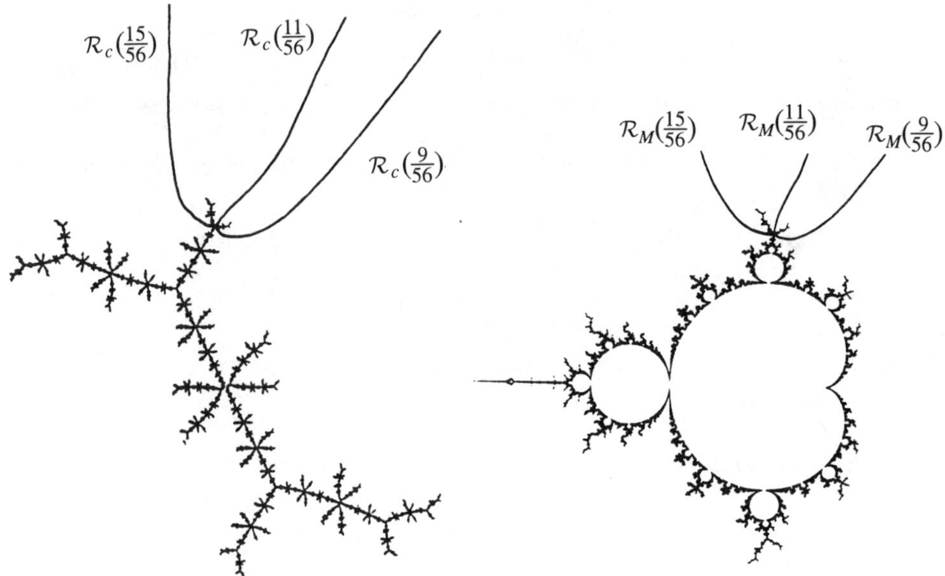

Figure 6. The Julia set of the quadratic polynomial f_c with $c \approx -0.101096 + i0.956287$. The critical point is eventually mapped onto the fixed point $\alpha_c = f_c^{\circ 4}(0)$. There are three external rays landing at c.

Figure 7. The boundary of the Mandelbrot set and the three external rays landing at $c \approx -0.101096 + i0.956287$.

Let $D : \mathbb{T} \to \mathbb{T}$ denote the doubling map $D(t) = 2t$. Note that under iteration by D the rationals with odd denominator are periodic and the rationals with even denominator are strictly preperiodic. The Figures 6 - 9 illustrate the relation between the landing properties in the dynamical plane and in the parameter plane formulated in Theorem 7.

Remark. The set of c-values for which the critical point is strictly preperiodic is dense in ∂M. In fact the smaller set consisting of c-values for which the critical point is eventually mapped onto a fixed point is dense in ∂M.

The rational rays which land at the c-values for which the critical point is eventually mapped onto the 'internal' fixed point are essential in the construction of the para-puzzles in the quadratic parameter plane.

3.1.2. The limbs and the wakes of the Mandelbrot set. The main hyperbolic component H_0 consists of the c-values for which the critical point is attracted to a fixed point under iteration. The boundary ∂H_0, the big cardioid, is parametrized by

$$\gamma_0(t) = \tfrac{1}{2}e^{2\pi it} - \tfrac{1}{4}e^{4\pi it}.$$

The point $\gamma_0(t)$ is the root point of a hyperbolic component exactly when t is rational and different from zero.

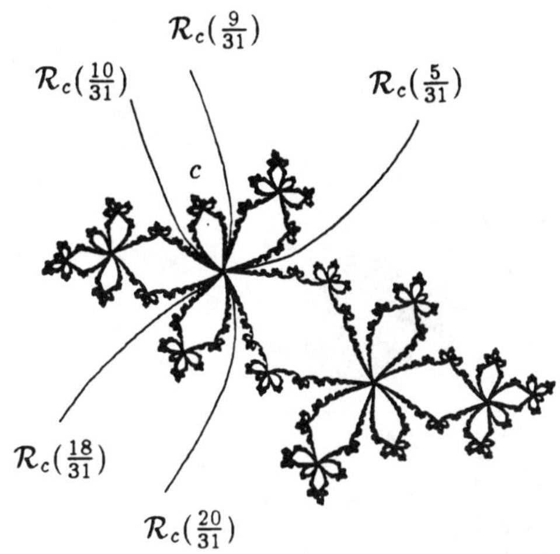

Figure 8. The Julia set of the quadratic polynomial f_c with $c = -0.504 + i0.568$ in the hyperbolic component H (shown to the right) and the five rays landing at the root point of the Fatou component containing c.

Figure 9. The boundary of the Mandelbrot set and the two external rays landing at r_H the root point of the hyperbolic component H.

Let $\frac{r}{s} \in \mathbb{Q} \cap \,]0,1[$ and $(r,s) = 1$. The $\frac{r}{s}$-*limb* $M_{r/s}$ of the Mandelbrot set M is the connected component of $M \setminus \overline{H_0}$ for which $\overline{M_{r/s}} \setminus M_{r/s} = \gamma_0(\frac{r}{s})$. Let $\theta^{\pm}_{r/s}$ denote the arguments of the two rational external rays which land at $\gamma_0(\frac{r}{s})$. The notation is chosen so that $0 < \theta^{-}_{r/s} < \theta^{+}_{r/s} < 1$.

Set

$$L(\tfrac{r}{s}) = R_M(\theta^{-}_{r/s}) \cup \{\gamma_0(\tfrac{r}{s})\} \cup R_M(\theta^{+}_{r/s}).$$

Then the $\frac{r}{s}$-*wake* $W_{r/s}$ is defined as the connected component of $\mathbb{C} \setminus L(\frac{r}{s})$ containing $M_{r/s}$. We shall give a characterization of polynomials belonging to a wake in the section on puzzles.

The $\frac{2}{5}$-limb of M and the $\frac{2}{5}$-wake are shown in Figure 9.

3.2. The parameter spaces of cubic polynomials.

For monic centered cubic polynomials $f_{(c_1,c_0)}(z) = z^3 + c_1 z + c_0$ the parameter space is $\mathbb{C}^2$. The topology of the parameter space is studied in [**BH1**] and [**BH2**]. The analogue of the Mandelbrot set is called the *Connectedness Locus* and denoted

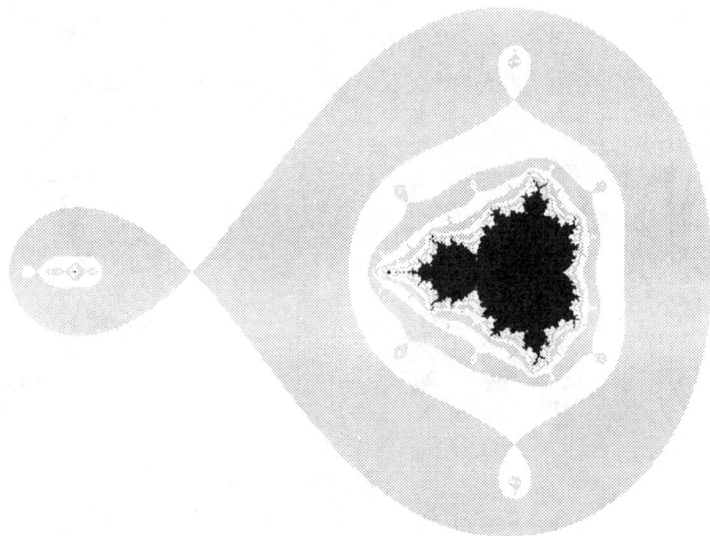

Figure 10. The set $B_\theta \subset \Lambda_\theta$ for $\theta = 0$ shown in black. The critical
graphs are seen as boundaries of gray or white regions.

$\mathcal{C}$; it equals the set of parameter values for which the Julia set of the corresponding
polynomial is connected. The complement of $\mathcal{C}$, the *Escape Locus*, is divided into
disjoint subsets $\mathcal{E}_1$ and $\mathcal{E}_2$ where $\mathcal{E}_j, j = 1, 2$, equals the set of polynomials for
which j critical points (counted with multiplicity) escape to infinity. The Julia sets
shown in Figures 1, 2 and 3 correspond to polynomials in $\mathcal{C}, \mathcal{E}_1$ and $\mathcal{E}_2$ respectively.

In this paper we are interested in cubic polynomials f with one critical point ω
escaping to infinity with escape rate $g_f(\omega) = \rho$ for some $\rho > 0$ and another critical
point ω_0 which escapes to infinity at a slower rate or not at all, i.e. $G(f) = \rho$
and $g_f(\omega_0) < \rho$. Let ω' denote the *co-critical point* to ω , i.e. the preimage of
the critical value $f(\omega)$ different from ω. Let θ denote the external argument
for which ω' belongs to the external ray $\mathcal{R}_f(\theta)$. Note that the external rays of
argument $\theta + \frac{j}{3}, j = 1, 2$, branch at the critical point ω. For any fixed ρ and θ
we define the parameter space

$$\Lambda_\theta = \Lambda_{\rho,\theta} = \{f \mid G(f) = g_f(\omega) = \rho \, , \omega' \in \mathcal{R}_f(\theta) \text{ and } g_f(\omega_0) < \rho\}.$$

The following theorem is proved in [**BH1**]:

THEOREM 8. *The parameter space* Λ_θ *is homeomorphic to an open disk.*

Let $B_\theta = B_{\rho,\theta}$ denote the set of polynomials in Λ_θ for which the critical orbit of
ω_0 is bounded. Set $h_\theta(f) = g_f(\omega_0)$. Then $h_\theta : \Lambda_\theta \to [0, \rho[$ is continuous on Λ_θ,
harmonic on $\Lambda_\theta \setminus B_\theta$ and equals zero on B_θ.

3.2.1. Critical graphs in the parameter space Λ_θ. One can prove that the critical points of h_θ in $\Lambda_\theta \setminus B_\theta$ are the polynomials f which satisfy a *critical orbit relation*, i.e. there exists an $n > 0$ such that $f^{\circ n}(\omega_0) = \omega$. The *critical graphs* are defined as the equipotentials which contain a critical point of h_θ. Hence a polynomial f belongs to a critical graph if and only if f satisfies a *critical potential relation*, i.e. there exists an $n > 0$ such that $g_f(f^{\circ n}(\omega_0)) = 3^n g_f(\omega_0) = g_f(\omega)$. The *critical potential values* are $\rho/3^n, n = 1, 2, \ldots$

The critical graphs are essential in the construction of the para-puzzles in the parameter space Λ_θ. (Compare with Figure 10.)

3.2.2. The Mañé-Sad-Sullivan decompositions of parameter spaces. For completeness we mention the first and the second Mañé-Sad-Sullivan decomposition of the quadratic parameter space $\mathbb{C}$ and the cubic parameter space $\mathbb{C}^2$. Any of the two parameter spaces is divided in two ways into disjoint subsets

$$\mathcal{R}_1 \sqcup \mathcal{S}_1 = \mathcal{R}_2 \sqcup \mathcal{S}_2$$

where $\mathcal{R}_j, j = 1, 2$, is open and dense in the parameter space.

For any connected component V_1 of $\mathcal{R}_1$ and for any two polynomials $f_1, f_2 \in V_1$ the polynomials f_1, f_2 are quasi-conformally conjugate on neighborhoods of the Julia sets.

For any connected component V_2 of $\mathcal{R}_2$ and for any two polynomials $f_1, f_2 \in V_2$ the polynomials f_1, f_2 are quasi-conformally conjugate in $\mathbb{C}$.

For quadratic polynomials the complement of $\mathcal{R}_j, j = 1, 2$, is given by

$$\mathcal{S}_1 = \partial M$$

$$\mathcal{S}_2 = \mathcal{S}_1 \cup \{f \mid f \text{ satisfies a c.o. relation }\}$$

where c.o. stands for 'critical orbit'. Note, that a quadratic polynomial satisfies a critical orbit relation if and only if the critical point is periodic.

For cubic polynomials the complement of $\mathcal{R}_j, j = 1, 2$, is given by

$$\mathcal{S}_1 = \partial \mathcal{C} \cup \partial \mathcal{E}_1$$

$$\mathcal{S}_2 = \mathcal{S}_1 \cup \{f \mid f \text{ satisfies a c.o. relation }\} \cup \{f \in \mathcal{E}_2 \mid f \text{ satisfies a c.p. relation }\}$$

where c.p. stands for 'critical potential'.

4. The method of divergence

As mentioned in the introduction the *method of divergence* is based on some classical results in Complex Analysis which we formulate in two proposition below.

Suppose $A \subset \mathbb{C}$ is an open subset which is doubly connected in $\widehat{\mathbb{C}}$, i.e. its complement in $\widehat{\mathbb{C}}$ has two connected components, one unbounded component containing ∞ and one bounded component; A is called an *annulus*. Any annulus is conformally equivalent to a *standard annulus* of the form

$$A_{r,R} = \{ z \in \mathbb{C} \mid r < |z| < R \} \quad \text{where} \quad 0 \leq r, \ R \leq \infty$$

for some r and R.

The *modulus* $\mathrm{mod}A$ of the annulus A is defined to be

$$\mathrm{mod}A \;=\; \tfrac{1}{2\pi}\log\tfrac{R}{r} \quad \text{if} \quad 0 < r \quad \text{and} \quad R < \infty$$

and $\mathrm{mod}A = \infty$ otherwise. The modulus is a conformal invariant.

Moreover, if $h : A_1 \to A_2$ is a covering map of degree k then $\mathrm{mod}A_1 = \tfrac{1}{k}\,\mathrm{mod}A_2$.

PROPOSITION 9 (GRÖTZSCH INEQUALITY). *Let $A \subset \mathbb{C}$ be an open annulus, and suppose (A_n) is a finite or infinite set of disjoint, open annuli $A_n \subset A$, each one winding around the bounded component of $\mathbb{C} \setminus A$. Then*

$$\sum_n \mathrm{mod}A_n \;\leq\; \mathrm{mod}A \;.$$

PROPOSITION 10. *If $A \subset \mathbb{C}$ is an open, bounded annulus with $\mathrm{mod}A = \infty$ then the bounded component of $\mathbb{C} \setminus A$ equals one single point.*

We can now formulate the *method of divergence* as follows:

COROLLARY 11. *Let $A \subset \mathbb{C}$ be an open, bounded annulus and suppose $(A_n)_{n \in \mathbb{N}}$ is an infinite set of disjoint, open annuli $A_n \subset A$, each one winding around the bounded component of $\mathbb{C} \setminus A$. If the series of moduli diverges*

$$\sum_{n=0}^{\infty} \mathrm{mod}A_n \;=\; \infty$$

then the bounded component of $\mathbb{C} \setminus A$ equals one single point.

Notice that no conclusion can be drawn if the series converges.

5. Puzzles

In this section we define the puzzle-pieces and the open annuli, first for a quadratic or a cubic polynomial with one unbounded critical orbit (i.e. with a disconnected Julia set) and then for a quadratic polynomial with one bounded critical orbit (i.e. with a connected Julia set). Moreover we prove – using the method of divergence – that the disconnected Julia set of a quadratic polynomial is totally disconnected.

5.1. The disconnected case.

Suppose f is either a quadratic polynomial with the critical point $\omega = 0$ escaping to ∞ (i.e. $f = f_c$ with $c \notin M$) or a cubic polynomial with one critical point ω escaping to ∞ and the other critical point ω_0 not escaping (i.e. $f \in \cup_{\rho,\theta} B_{\rho,\theta}$). Recall that $G(f)$ denotes the maximal escape rate of the critical points. The set

$$P_0 \;=\; \{z \in \mathbb{C} \mid g_f(z) \leq d \cdot G(f)\}$$

is a closed topological disk (d equals 2 when f is quadratic and 3 when f is cubic). Compare with Figures 4 and 5.

The set K_f is disconnected and plays the role of the set X mentioned in the introduction. Define for any $x \in K_f$ and any $n \in \mathbb{N}$ the set $P_n(x)$ to be the closure
of the connected component containing x of the set

$$f^{-n}(\overset{\circ}{P}_0) = \{z \in \mathbb{C} \mid g_f(z) < \tfrac{1}{d^{n-1}}\, G(f)\}.$$

Each $P_n(x)$ is a closed topological disk. The disk $P_n(x)$ is called a *puzzle-piece*
of depth n. The collection of such puzzle-pieces forms *the puzzle* of depth n. The
puzzle of depth $n > 0$ is bounded by the critical graph

$$\Gamma_f(n) = \{z \in \mathbb{C} \mid g_f(z) = \tfrac{1}{d^{n-1}}\, G(f)\}.$$

The puzzle-piece $C_n = P_n(\omega_0)$ is called the *critical piece* of depth n and $V_{n-1} =
P_{n-1}(f(\omega_0))$ is called the *critical value piece* of depth $n - 1$.

Consider the map $f|_{P_n(x)} : P_n(x) \to P_{n-1}(f(x))$. It is always a homeomorphism
in the quadratic case. In the cubic case it is a homeomorphism if $P_n(x) \neq C_n$ and
a ramified covering of degree 2 if $P_n(x) = C_n$.

For each $x \in K_f$ we denote the connected component of K_f containing x by
$K_f(x)$. Each connected component is specified by the nested sequence of puzzle-
pieces

$$P_0 \supset P_1(x) \supset \cdots \supset P_n(x) \supset \ldots$$

and

$$K_f(x) = \bigcap_{n=0}^{\infty} P_n(x).$$

The nested sequence of puzzle-pieces surrounding $K_f(x)$ is mapped by f to the
nested sequence surrounding $K_f(f(x))$ and

$$f(K_f(x)) = K_f(f(x)).$$

Define the open annulus

$$A_0 = \{z \in \mathbb{C} \mid G(f) < g_f(z) < d \cdot G(f)\}$$

and note that φ_f maps A_0 isomorphically onto the standard annulus $A_{r,R}$ with
$r = \exp(G(f))$ and $R = \exp(d \cdot G(f))$. It follows that $\mathrm{mod} A_0 = \tfrac{1}{2\pi}\,(d-1)G(f)$.
Define for each $x \in K_f$ and each $n \in \mathbb{N}$ the open annulus

$$A_n(x) = P_n(x) \cap f^{-n}(A_0) .$$

The annulus is contained in $P_n(x) \setminus P_{n+1}(x)$ and winds around $K_f(x)$. The open,
bounded annulus $A(x) = \overset{\circ}{P}_0 \setminus K_f(x) \Subset \mathbb{C}$ contains an infinity of disjoint open
annuli $A_n(x)$, $n \geq 0$, all winding around $K_f(x)$.

The map $f|_{A_n(x)} : A_n(x) \to A_{n-1}(f(x))$, $n > 0$, is always a conformal equivalence
in the quadratic case. In the cubic case it is also a conformal equivalence if $P_n(x) \neq$

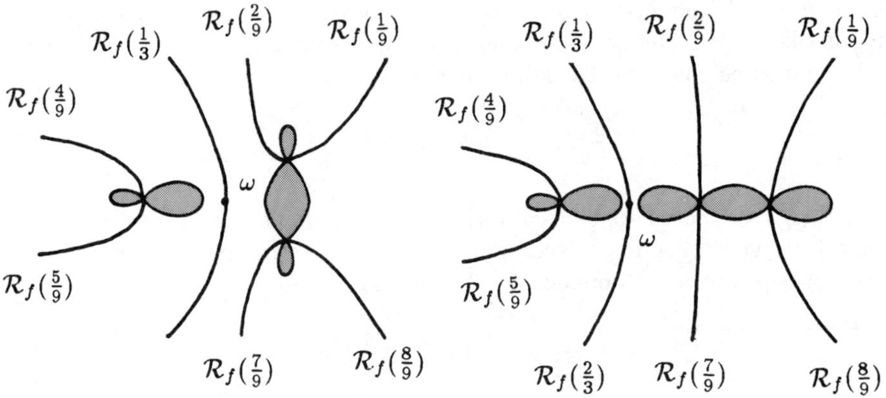

Figure 11. The two possible cubic puzzles of depth 2 together with the external rays branching at ω and $z \in f^{-1}(\omega)$. (The external arguments fit with $f \in \Lambda_\theta$ for $\theta = 0$.) To the left $V_1 = C_1$ and to the right $V_1 \neq C_1$.

C_n and a covering map of degree 2 if $P_n(x) = C_n$. We call the annulus $A_n(\omega_0)$ *critical*. Note that

$$\mathrm{mod}A_n(x) \doteq \begin{cases} \mathrm{mod}A_{n-1}(f(x)) & \text{if} \quad A_n(x) \quad \text{is not critical} \\ \frac{1}{2}\,\mathrm{mod}A_{n-1}(f(x)) & \text{if} \quad A_n(x) \quad \text{is critical.} \end{cases}$$

In the quadratic case all the annuli have the same modulus, equal to $\mathrm{mod}A_0$, and all the puzzles of depth n are topologically the same.

The cubic case is much richer. Not only do the annuli have different moduli. The main reason for the richness is that there is an infinite tree of qualitatively different puzzles. In order to draw the puzzle of depth $n > 1$ for a given cubic polynomial f we must specify the nested sequence of critical value pieces

$$P_0 = V_0 \supset V_1 \cdots \supset V_{n-1}.$$

The different puzzles correspond to qualitatively different behavior of the critical point ω_0. There are two different puzzles of depth 2 for an arbitrary θ. We illustrate this in Figure 11 for $\theta = 0$. Note that there are five different puzzles of depth 3 since there are three respectively two different possible choices for $V_2 \subset V_1$. The critical graphs look the same for all θ, but the arguments of the rays branching at the critical points of g_f depends on θ.

Note that all the annuli are disjoint from K_f.

5.1.1. The total disconnectedness in the quadratic case. Suppose f_c is a quadratic polynomial with $c \notin M$.

The classical proof of the fact that K_c is a Cantor set applies the Poincaré metric on the open topological disk equal to the interior of P_0. The inverse of the map

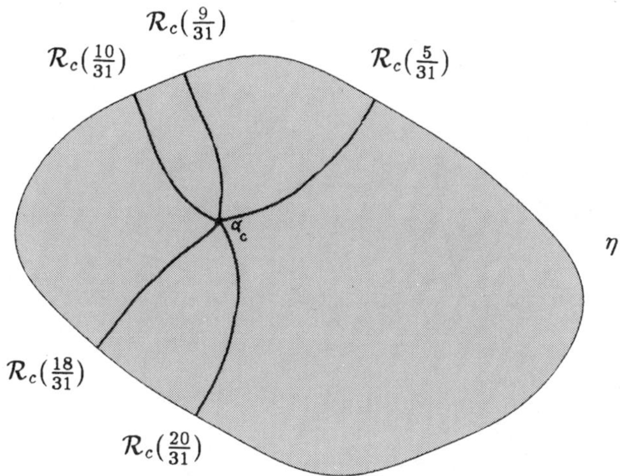

Figure 12. The 5-star for $c \in W_{2/5}$ and the puzzle of depth 0.

$f_c|_{P_1} : P_1 \rightarrow P_0$, restricted to the interior, is a strong contraction in the Poincaré metric for any of the two puzzle-pieces P_1 of depth 1. Any infinite composition of the two inverse maps defines a constant map from $\overset{\circ}{P}_0$ and the total disconnectedness follows.

From the construction of puzzles and annuli we obtain an alternative proof of the total disconnectedness based on the method of divergence: For each $x \in K_c$ the open, bounded annulus $A(x) = \overset{\circ}{P}_0 \setminus K_c(x)$ contains an infinity of disjoint open annuli $A_n(x)$, all winding around $K_c(x)$ and all with the same modulus. It follows from Corollary 11 that $K_c(x)$ is equal to the single point $\{x\}$. Hence $K_c = J_c$ is totally disconnected.

Remark. The method of divergence can also be used to prove that K_f is totally disconnected for any polynomial f with all critical points escaping to ∞. The relevant annuli to consider are those between critical graphs of potential less than or equal to $\min_{\omega \in \Omega(f)} g_f(\omega)$. Each annulus has a modulus which equals a positive number, belonging to a finite list of possibilities. This ends the comments on the proof of Theorem 1.

5.2. The connected case.

We start by characterizing the polynomials belonging to one of the wakes, say the $\frac{r}{s}$-wake $W_{r/s}$. Each polynomial f_c with $c \in W_{r/s}$ has two repelling fixed points. One fixed point is the landing point of the external 0-ray, to be denoted β_c. The other fixed point, to be denoted α_c, is the landing point of a cycle of s rational external rays. The set of external arguments $\mathrm{Arg}_{r/s}$ of these s rays is constant through out the $\frac{r}{s}$-wake and characterizes the wake. Each of these s rays is mapped onto the r-th ray counted counterclockwise in the cyclic order. The s-star formed by these s rays together with their common landing point α_c play an important role in the construction of puzzle-pieces. The complement in $\mathbb{C}$ has s connected components; the critical value c belongs to one of them. The external arguments of the two rays

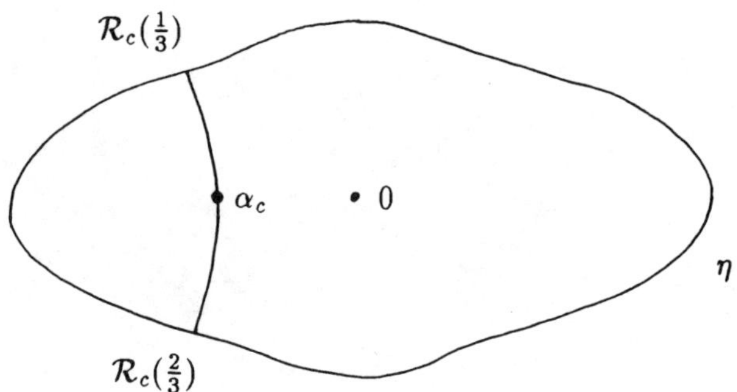

Figure 13. The graph $\Gamma_c(0)$ for $c \in M_{1/2}$.

adjacent to the critical value component are constant through out the wake. The values are $\theta^{\pm}_{r/s}$. Note that these arguments are the same as the arguments in the parameter plane of the rays bounding the $\frac{r}{s}$-wake. The set $\mathrm{Arg}_{r/s}$ can be expressed by $\theta^{\pm}_{r/s}$ as follows

$$\mathrm{Arg}_{r/s} = \{2^j \theta^-_{r/s} \mid j = 0, 1, \ldots, s-1\} = \{2^j \theta^+_{r/s} \mid j = 0, 1, \ldots, s-1\}.$$

For instance we have $\theta^-_{1/2} = \frac{1}{3}$, $\theta^-_{1/3} = \frac{1}{7}$ and $\theta^-_{2/5} = \frac{9}{31}$.

Figure 12 shows the 5-star, the arguments $\theta^{\pm}(\frac{2}{5})$, the set $\mathrm{Arg}_{2/5}$ and the puzzle of depth 0.

Set

$$F_M = \bigcup_{n>0}^{\infty} \{c \mid f_c^{on}(c) = \alpha_c\}.$$

If $c \in F_M$ then the critical point is strictly preperiodic and the Julia set J_c is classically known to be locally connected. It suffices to consider c-values in the disconnected subset

$$M^*_{r/s} = M_{r/s} \setminus F_M$$

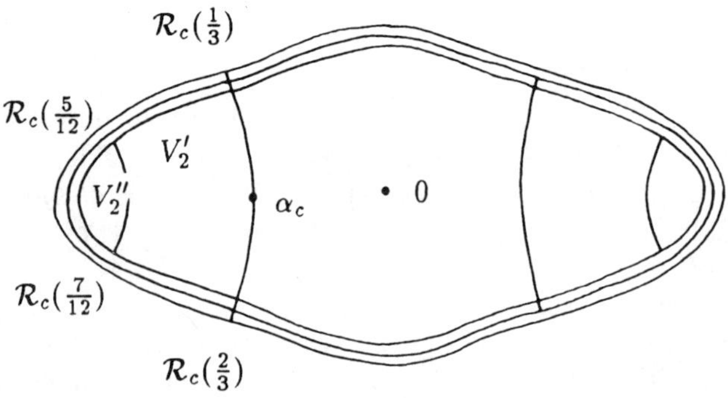

Figure 14. The graphs $\Gamma_c(n)$, $n = 0, 1, 2$, for $c \in M_{1/2}$.

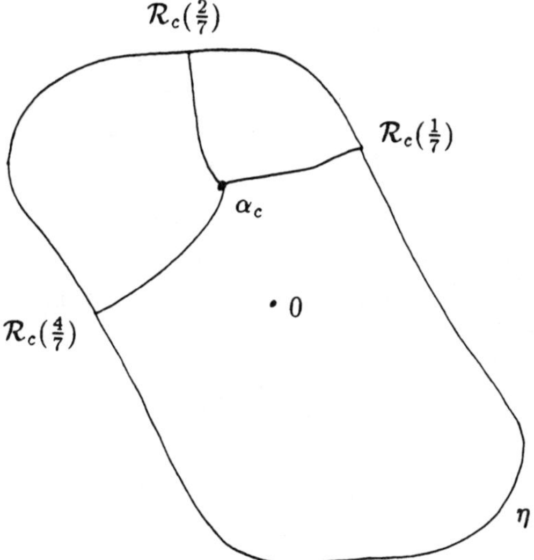

Figure 15. The graph $\Gamma_c(0)$ for $c \in M_{1/3}$.

of the $\frac{r}{s}$-limb. In the following we assume $c \in M^*_{r/s}$.

Choose any potential value $\eta > 0$ and consider the closed topological disks

$$\Delta_c(n) = \{z \in \mathbb{C} \mid g_f(z) \leq \tfrac{\eta}{2^n}\}.$$

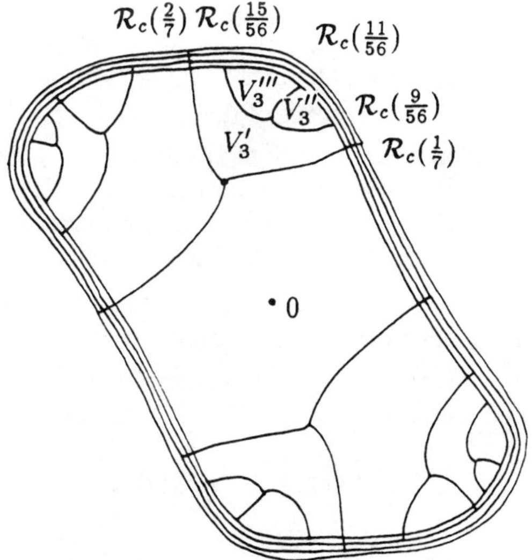

Figure 16. The graphs $\Gamma_c(n)$, $n = 0, 1, 2, 3$, for $c \in M_{1/3}$.

Define the s-star restricted to $\Delta_c(0)$ as

$$S_c(\tfrac{r}{s}) = \{\alpha_c\} \cup \bigcup_{\theta \in \, \mathrm{Arg}_{r/s}} (\mathcal{R}_c(\theta) \cap \Delta_c(0)).$$

Define the graph

$$\Gamma_c(0) = S_c(\tfrac{r}{s}) \cup \partial\Delta_c(0)).$$

Compare with Figures 13 and 15.

The s puzzle-pieces of depth 0 are defined to be the closure of the connected components of $\Delta_c(0) \setminus \Gamma_c(0)$. Define the graphs $\Gamma_c(n)$ as the preimages of $\Gamma_c(0)$

$$\Gamma_c(n) = f_c^{-n}(\Gamma_c(0)).$$

The puzzle-pieces of depth n are defined to be the closure of the connected components of $\Delta_c(n) \setminus \Gamma_c(n)$ and the collection of such puzzle-pieces is called *the puzzle of depth n*.

Notice that the external rays which are part of the graph $\Gamma_c(n)$ have arguments in the set $D^{-n}(\mathrm{Arg}_{r/s})$, independent of c. However the way the set $D^{-n}(\mathrm{Arg}_{r/s})$ is divided into subsets, each consisting of s arguments associated to one of the preimages $y \in f_c^{-(n+1)}(\alpha_c)$, depends on c. In the section on quadratic para-puzzles we shall formulate this dependence precisely.

The graphs $\Gamma_c(n)$ replace the critical graphs $\Gamma_f(n)$ in the disconnected case; in both cases the puzzle-pieces are the closure of the bounded connected components of the complement of the graphs.

For $c \in M_{r/s}^*$ the filled Julia set K_c is connected. The disconnected set which plays the role of the set X mentioned in the introduction is the set

$$K_c^* = K_c \setminus F_c \quad \text{where} \quad F_c = \bigcup_{n=0}^{\infty} f_c^{-n}(\alpha_c).$$

Choose any $x \in K_c^*$ and let $P_n(x)$ denote the puzzle-piece of depth n containing x. The puzzle-piece $C_n = P_n(0)$ is called the *critical piece* of depth n and $V_{n-1} = P_{n-1}(c)$ is called the *critical value piece* of depth $n-1$.

Since we want to prove local connectivity of the Julia set J_c for certain c-values it is important to observe that the set $K_c \cap P_n(x)$ is connected for all $x \in K_c^*$ and for all $n \geq 0$. This is proved first for $n = 0$ and then by induction in n. Moreover, if $y \in f_c^{-n}(\alpha_c)$ then $K_c \cap \bigcup_{y \in P_k(x)} P_k(x)$ is connected.

Consider the map $f_c|_{P_n(x)} : P_n(x) \to P_{n-1}(f_c(x))$. The map is a homeomorphism if $P_n(x) \neq C_n$ and a ramified covering of degree 2 if $P_n(x) = C_n$.

For each $x \in K_c^*$ we denote the connected component of K_c^* containing x by $K_c^*(x)$. Each connected component is specified by the nested sequence of puzzle-pieces

$$P_0(x) \supset P_1(x) \supset \cdots \supset P_n(x) \supset \ldots$$

and

$$K_c^*(x) = \bigcap_{n=0}^{\infty} P_n(x).$$

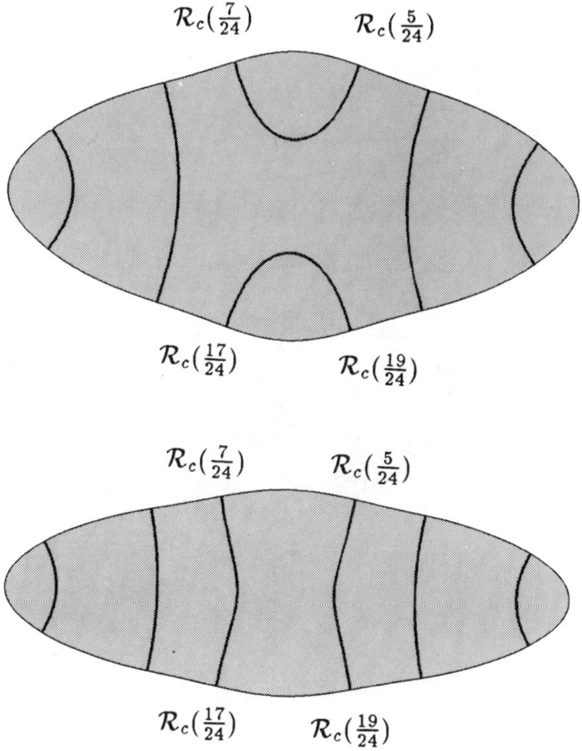

Figure 17. The two possible puzzles of depth 3 for $c \in M_{1/2}$ corresponding to V_2 equal to V_2' above and V_2'' below (compare with Figure 14).

The nested sequence of puzzle-pieces surrounding $K_c^*(x)$ is mapped by f_c to the nested sequence surrounding $K_f^*(f(x))$ and

$$f_c(K_c^*(x)) = K_c^*(f_c(x)).$$

In order to define the puzzle of depth n for a given quadratic polynomial we must specify the nested sequence of critical value pieces

$$V_0 \supset V_1 \supset \cdots \supset V_{n-1}.$$

Note that for a given $\frac{r}{s}$ we have

$$V_0 \cap K_c = V_1 \cap K_c = \cdots = V_{s-1} \cap K_c$$

but that V_{s-1} is subdivided into s pieces at depth s. Each piece can be the critical value piece for certain polynomials (we shall be more precise in the section on quadratic para-puzzles) and each possibility gives rise to a different puzzle of depth s. This is illustrated in Figure 17 for $\frac{r}{s} = \frac{1}{2}$ and in Figure 18 for $\frac{r}{s} = \frac{1}{3}$. At a deeper

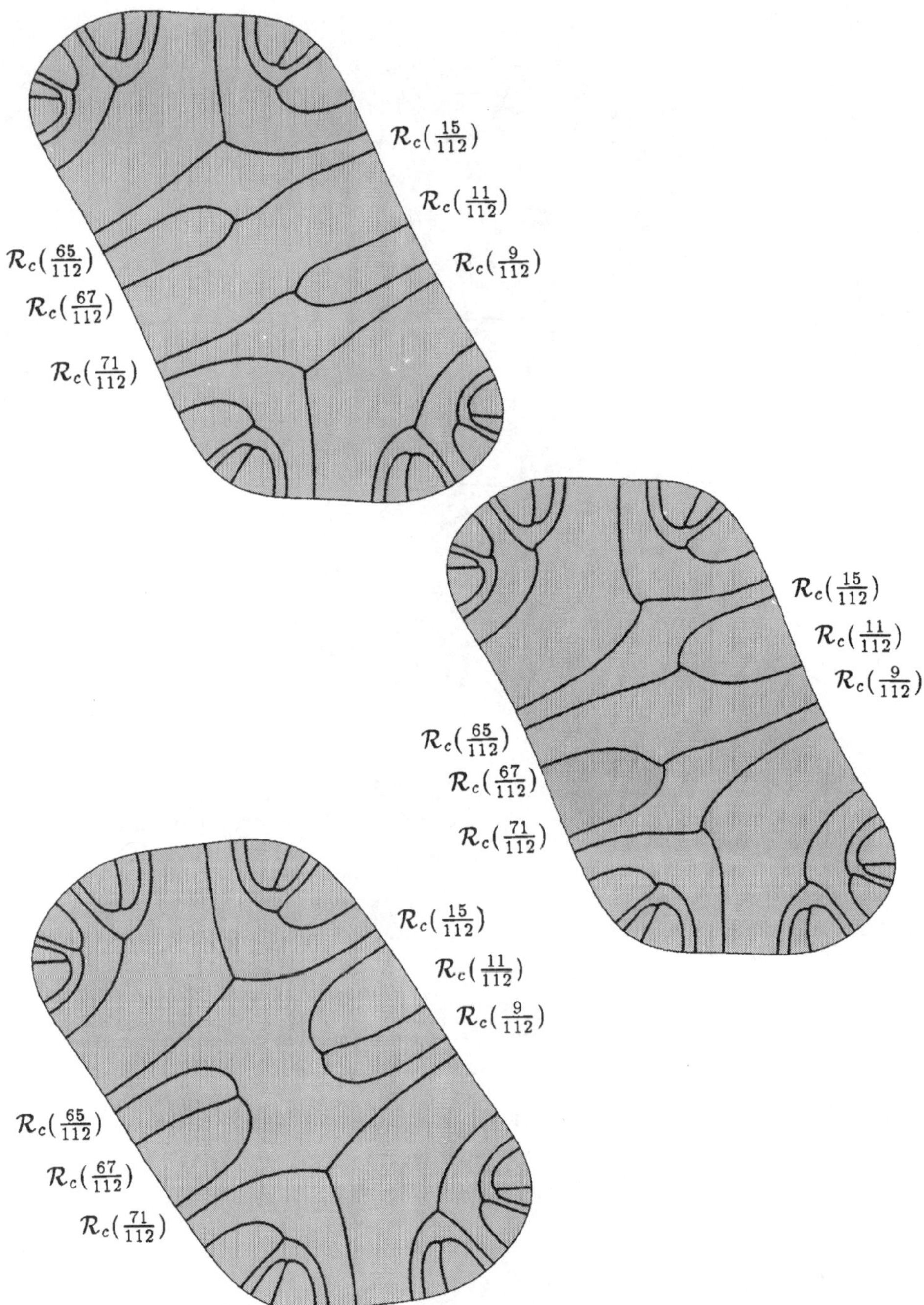

Figure 18. The three possible puzzles of depth 4 for $c \in M_{1/3}$ corresponding to V_3 equal to V_3', V_3'', V_3''' respectively (compare with Figure 16).

level the critical value piece is again subdivided, etc., etc. Hence there is an infinite tree of different puzzles.

Compare also with Colored Plates 3 and 4. Plate 3 shows the filled Julia set K_c for a c-value in $M_{1/3}$ with $V_3 = V_3'$ together with the external rays defining the puzzle of depth 4. Plate 4 is a blow up around the critical point of Plate 3 together with the external rays defining the puzzle of depth 7.

Let $x \in K_c^*$ and define

$$A_n(x) = \overset{\circ}{P}_n(x) \setminus P_{n+1}(x).$$

If $\partial P_n(x) \cap \partial P_{n+1}(x) \neq \emptyset$ then $A_n(x)$ does not form an annulus; we call it a *degenerate annulus*, and define its modulus $\mathrm{mod}A_n(x)$ to be zero. If $\partial P_n(x) \cap \partial P_{n+1}(x) = \emptyset$ then $A_n(x)$ is an ordinary annulus with a positive modulus.

The map $f_c|_{A_n(x)} : A_n(x) \to A_{n-1}(f_c(x))$ is a covering map of degree 2 if $P_{n+1}(x) = C_{n+1}$ and is a holomorphic isomorphism if $P_n(x) \neq C_n$. The annulus $A_n(0)$ is called *critical*. The annulus $A_n(x)$ is called *semi-critical* if $0 \in A_n(x)$; in this case $A_n(x)$ is not mapped onto $A_{n-1}(f_c(x))$ but contains this set as a proper subset. Compare for instance Figures 14 and 17. In the case where $V_3 = V_3'$ the annulus $A_2 \subset C_2 \setminus P_3$ with $P_3 \neq C_3$ is semi-critical and the annulus $A_2 \subset C_2 \setminus C_3$ is degenerate. While in the case where $V_3 = V_3''$ the annulus $A_2 \subset C_2 \setminus C_3$ is critical and an ordinary annulus. Note that

$$\mathrm{mod}A_n(x) = \begin{cases} \mathrm{mod}A_{n-1}(f_c(x)) & \text{if} \quad A_n(x) \quad \text{is not (semi-)critical} \\ \frac{1}{2}\,\mathrm{mod}A_{n-1}(f_c(x)) & \text{if} \quad A_n(x) \quad \text{is critical.} \end{cases}$$

Note that in the connected quadratic case many of the annuli intersect the filled Julia set K_c.

6. Para-puzzles

There is a close relation between the infinite tree of puzzles and the para-puzzle-pieces we define in this section. We start with quadratic polynomials in $M_{r/s}^*$ for any fixed $\frac{r}{s}$ and continue with cubic polynomials in B_θ for any fixed $\theta \in \mathbb{T}$.

6.1. Quadratic para-puzzles.

Consider polynomials in a fixed wake $W_{r/s}$. Fix as before a potential value $\eta > 0$. Set

$$\Delta_M(n) = \{c \in \overline{W_{r/s}} \mid G_M(c) \leq \tfrac{\eta}{2^n}\}$$

and define the graph

$$\Gamma_M(0) = \partial \Delta_M(0).$$

Figure 19 shows the graph $\Gamma_M(0)$ for $\frac{r}{s} = \frac{1}{2}$ and $\frac{r}{s} = \frac{1}{3}$. The *para-puzzle-piece* MP_0 of depth 0 equals $\Delta_M(0)$.

Consider for a given $n > 0$ the set of points

$$F_M(n) \cap M_{r/s} = \{c \in M_{r/s} \mid f_c^{\circ n}(0) = \alpha_c\}.$$

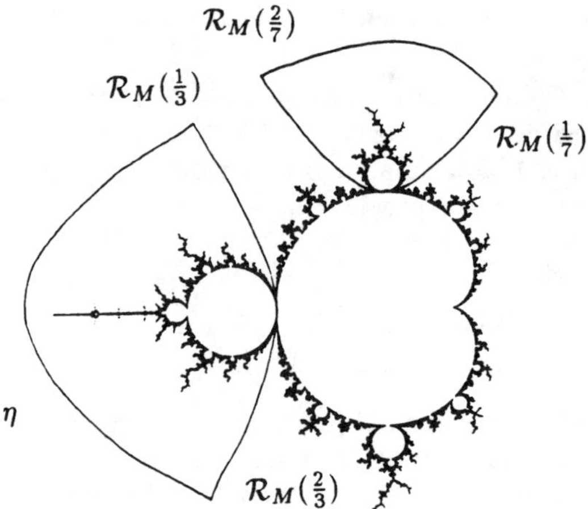

Figure 19. The two graphs $\Gamma_M(0)$ for $\tfrac{r}{s} = \tfrac{1}{2}$ and $\tfrac{r}{s} = \tfrac{1}{3}$.

Each c in this set is the landing point of s external rays. Let $\Gamma_M(n)$ denote the graph contained in $\Delta_M(n)$ and consisting of the boundary of this region together with the points in $F_M(n) \cap M_{r/s}$ and the rays landing at these points. Compare with Figure 20 which illustrates $\Gamma_M(3)$ for $\tfrac{r}{s} = \tfrac{1}{2}$ and $\Gamma_M(4)$ for $\tfrac{r}{s} = \tfrac{1}{3}$. A para-puzzle piece of depth n is defined to be the closure of a connected component of $\Delta_M(n) \setminus \Gamma_M(n)$ and the collection of such para-puzzle-pieces is called *the para-puzzle* of depth n. One can prove that all the para-puzzle-pieces intersect $M_{r/s}$ in a connected set.

The set $M_{r/s}^* = M_{r/s} \setminus F_M$ is disconnected and plays the role of the set X mentioned in the introduction. For each $c \in M_{r/s}^*$ we denote the connected component of $M_{r/s}^*$ containing c by $M_{r/s}^*(c)$. Each connected component is specified by the

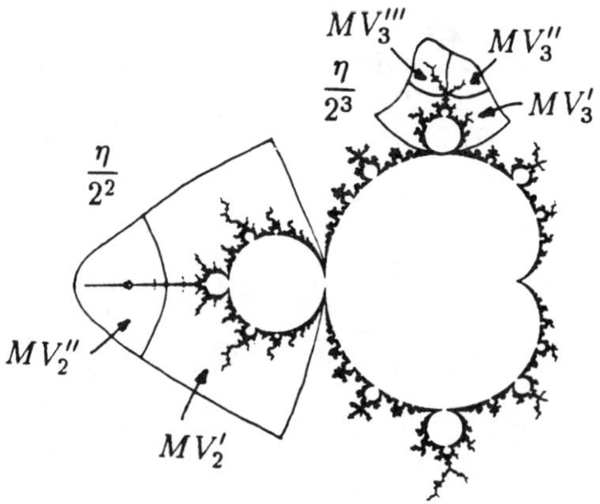

Figure 20. The para-puzzles of depth 2 and 3 for $\tfrac{r}{s} = \tfrac{1}{2}$ and $\tfrac{1}{3}$.

nested sequence of para-puzzle-pieces

$$MP_0(c) \supset MP_1(c) \supset \cdots \supset MP_n(c) \supset \ldots$$

and

$$M^*_{r/s}(c) = \bigcap_{n=0}^{\infty} MP_n(c).$$

For each depth n the para-puzzle defines a division of parameter space which is re-lated to the different puzzles of depth n in the following way. For all c in the interior of a para-puzzle-piece MP_n the graphs $\Gamma_c(n)$ in the dynamical plane are 'the same', i.e. the s external arguments associated to any of the preimages $y \in f_c^{-(n+1)}(\alpha_c)$ are the same. Moreover the boundary of MP_n is 'the same'as the boundary of the critical value piece ∂V_n for any $c \in MP_n$. Compare Figure 20 with Figures 17 and 18. The two respectively three para-puzzle-pieces in Figure 20 correspond to the two different puzzles of depth 3 respectively three different puzzles of depth 4. In Figure 20 we have denoted the para-puzzle-pieces by MV_2', MV_2'' respectively MV_3', MV_3'', MV_3''' in order to emphazise the connection with the different choices of the critical value piece at depth 2 respectively 3 illustrated in Figures 14 and 16.

For each $c \in M^*_{r/s}$ we define

$$MA_n(c) = \overset{\circ}{MP}_n(c) \setminus MP_{n+1}(c).$$

As in the dynamical plane there are degenerate annuli with zero modulus and ordinary annuli with positive modulus. Note that many of the annuli MA_n intersect the Mandelbrot set.

6.2. Cubic para-puzzles.

Consider cubic polynomials in $B_\theta \subset \Lambda_\theta$ for a fixed θ and a fixed potential ρ. In order to emphazise the dependence of the parameter $\lambda \in \Lambda_\theta$ we shall denote the corresponding cubic polynomial by f_λ and use the notation K_λ etc.

The para-puzzle of depth 0 has one piece, namely the closure of Λ_θ. The para-puzzle-pieces of depth n are defined as the closure of the connected components of

$$\Lambda_\theta(n) = \{\lambda \in \Lambda_\theta \mid h_\theta(\lambda) < \tfrac{\rho}{3^n}\}.$$

We denote the *para-puzzle-piece* of depth n containing $\lambda \in B_\theta$ by $BP_n(\lambda)$. The set of para-puzzle-pieces of depth n is the *para-puzzle* of depth n. The para-puzzle of depth $n > 0$ is bounded by the critical graph of potential $\rho/3^n$.

The set B_θ is disconnected and plays the role of the set X mentioned in the intro-duction. For each $\lambda \in B_\theta$ we denote the connected component of B_θ containing λ by $B_\theta(\lambda)$. Each connected component is specified by the nested sequence of para-puzzle-pieces

$$BP_0(\lambda) \supset BP_1(\lambda) \supset \cdots \supset BP_n(\lambda) \supset \ldots$$

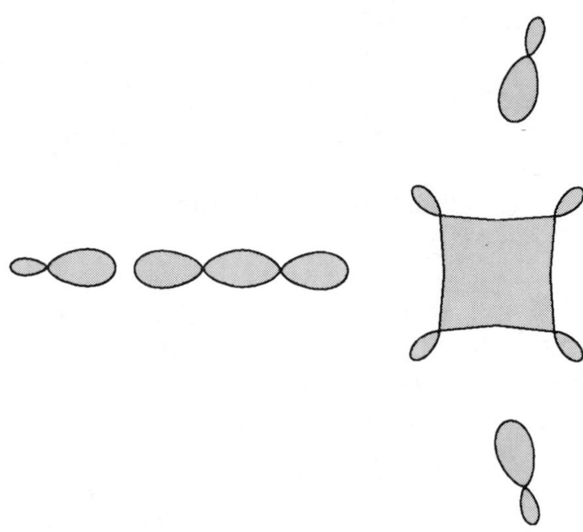

Figure 21. The cubic para-puzzle of depth 3.

and

$$B_\theta(\lambda) = \bigcap_{n=0}^{\infty} BP_n(\lambda).$$

For each depth n the para-puzzle defines a division of parameter space Λ_θ which is related to the different puzzles in the following way. For all μ in a para-puzzle-piece $BP_n(\lambda)$ the critical graph $\Gamma_\mu(n)$ in the dynamical plane is 'the same'. Moreover the boundary of $BP_n(\lambda)$ is 'the same' as the boundary of the critical value piece ∂V_n for any $\mu \in BP_n(\lambda)$. Compare Figure 21 with Figures 5 and 11. The infinite tree of puzzles is reflected in the para-puzzles at deeper and deeper levels.

Let $\lambda \in B_\theta$ and define

$$BA_n(\lambda) = \{\mu \in BP_n(\lambda) \mid \tfrac{\rho}{3^n} < h_\theta(\mu) < \tfrac{\rho}{3^{n-1}}\}.$$

Note that all the annuli are disjoint from B_θ. Compare with Figure 10 and Colored Plates 1 and 2.

The similarity between the partition in the parameter space and the partition in the dynamical planes is more pronounced in the cubic case than in the quadratic case. One can prove ([**BH2**]) that the annulus $BA_n(\lambda)$ for any $\mu \in BP_n(\lambda)$ is conformally equivalent to the critical value annulus $A_n(f_\mu(\omega_0))$. Therefore

$$\sum_{n=0}^{\infty} \mathrm{mod} BA_n(\lambda) = \sum_{n=0}^{\infty} \mathrm{mod} A_n(f_\lambda(\omega_0)),$$

and $B_\theta(\lambda)$ is a point component in B_θ if and only if $K_\lambda(f_\lambda(\omega_0))$ is a point component in K_λ.

The similarity between the partition in parameter space and the partition in the dynamical planes cannot be that simple in the quadratic case. If an annulus $MA_n(c)$ contains part of M then the corresponding critical value annulus in the dynamical plane contains part of the filled Julia set. There is no obvious map from the M-part of the annulus to the K_c-part of the corresponding annulus and hence no reason for the annuli to be conformally equivalent. It is therefore much more technical in the quadratic case to prove the statement which will lead to the result analogous to the one above. The statement is that the series $\sum_{n=0}^{\infty} \mathrm{mod} M A_n(c)$ is divergent if and only if the series $\sum_{n=0}^{\infty} \mathrm{mod} A_n(c)$ is divergent. The analogous result is that $M_{r/s}^*(c)$ is a point component in $M_{r/s}^*$ if and only if $K_c^*(c)$ is a point component in K_c^*.

7. The main Theorems

We are now ready to formulate the main theorems both in the dynamical plane and in the parameter space and both in the cubic and in the quadratic case.

7.1. Theorems in the cubic case.

Let f be a cubic polynomial with the critical orbit of ω_0 bounded and the other critical orbit unbounded. We call the critical component $K_f(\omega_0)$ *periodic* if there exists a $k > 0$ so that $f^{\circ k}(K_f(\omega_0)) = K_f(\omega_0)$. Note that if the critical component is not periodic then the post-critical orbit $\{f^{\circ n}(\omega_0)\}_{n>0}$ does not intersect the critical component $K_f(\omega_0)$.

THEOREM 12 DYNAMICAL THEOREM FOR CUBIC POLYNOMIALS. *Let $\lambda \in B_{\rho,\theta}$ for some $\rho > 0$ and some $\theta \in \mathbb{T}$. Then the filled Julia set K_λ is a Cantor set if and only if the critical component $K_\lambda(\omega_0)$ is not periodic.*

THEOREM 13 PARAMETER SPACE THEOREM FOR CUBIC POLYNOMIALS. *Let $\lambda \in B_{\rho,\theta}$ for some $\rho > 0$ and some $\theta \in \mathbb{T}$. Then the connected component $B_{\rho,\theta}(\lambda)$ is a point component of B_θ if and only if the filled Julia set K_λ is a Cantor set.*

Remarks. We shall not discuss the other possibilities for K_λ and $B_{\rho,\theta}$ in this paper but only mention the results:
(1) Let $\lambda \in B_{\rho,\theta}$ for some $\rho > 0$ and some $\theta \in \mathbb{T}$. Suppose the critical component $K_\lambda(\omega_0)$ is periodic of period k.
(a) Then there exist an $n_0 \geq k$ and a unique c in the Mandelbrot set such that

$$h_n = f^{\circ k}|_{C_n} : C_n \to C_{n-k}$$

is quadratic-like for all $n \geq n_0$ and h_n is hybrid equivalent to the quadratic polynomial $f_c(z) = z^2 + c$. In particular, $K_\lambda(\omega_0)$ is homeomorphic to K_c.
(b) A connected component $K_\lambda(x)$ of the filled Julia set is homeomorphic to the critical component $K_\lambda(\omega_0)$ if and only if it is a preimage of the critical component under some iterate of f_λ. All other connected components are point components.
(c) The connected component $B_{\rho,\theta}(\lambda)$ of $B_{\rho,\theta}$ in the parameter space is a homeomorphic copy of the Mandelbrot set.
(2) Figure 2 illustrates (a) and (b) and Figure 10 illustrates (c).
(3) The filled Julia set K_λ has uncountably many connected components, countably many components which are not point components and uncountably many

point components. The set $B_{\rho,\theta}$ has also uncountably many connected components, countably many components which are copies of the Mandelbrot set and uncountably many point components.

(4) Note that the *only-if* part of Theorem 12 and Theorem 13 follows from (a) and (c) above.

7.2. Theorems in the quadratic case.

Let $c \in M \setminus (\overline{H_0} \cup F_M)$, i.e. the polynomial f_c has two repelling fixed points and the critical orbit does not contain the fixed point α_c. Similarly to the cubic case we call the critical component $K_c^*(0)$ *periodic* if there exists a $k > 0$ so that $f_c^{\circ k}(K_c^*(0)) = K_c^*(0)$.

THEOREM 14 DYNAMICAL THEOREM FOR QUADRATIC POLYNOMIALS. *Let $c \in M \setminus (\overline{H_0} \cup F_M)$. The disconnected set K_c^* is totally disconnected if and only if the critical component $K_c^*(0)$ is not periodic. In that case the filled Julia set K_c is locally connected.*

THEOREM 15 PARAMETER PLANE THEOREM FOR QUADRATIC POLYNOMIALS. *Let $c \in M_{r/s}^*$ for some $\frac{r}{s}$. The connected component $M_{r/s}^*(c)$ is a point component of $M_{r/s}^*$ if and only if K_c^* is totally disconnected. In that case the Mandelbrot set M is locally connected at c.*

Remarks. We shall not discuss the other possibilities in this paper but only mention the known results (see [**H**]).

(1) Let c be as in Theorem 14 and suppose $c \in M_{r/s}^*$. If the critical component $K_c^*(0)$ is periodic then f_c is *renormalizable* and the closure of the connected component $M_{r/s}^*(c)$ is a homeomorphic copy of the Mandelbrot set.

(2) The two theorems above can be extended to quadratic polynomials f_c of the following type: f_c is *finitely often renormalizable*, all periodic orbits are repelling and the critical point is not strictly preperiodic. For such c-values K_c is locally connected, furthermore M is locally connected at c.

(3) Theorem 14 and 15 show examples where local connectivity of a quadratic Julia set is linked to local connectivity of the Mandelbrot set at the corresponding parameter value. Note that this is not a general principle: there are examples of quadratic polynomials f_c where J_c is not locally connected, but M is locally connected at c. The Cremer polynomials on the boundary of hyperbolic components are such examples.

(4) Recently it has also been proved that some *infinitely often renormalizable* quadratic polynomials have locally connected Julia set, see [**HJ**], [**J**] and [**L**]. The technique is to construct divisions similar to puzzles and to use the method of divergence. But there are also examples of infinitely often renormalizable quadratic polynomials with non-locally connected Julia set (see [**M**]) and even examples for which the Julia set is not arc-wise connected (see [**S**]).

All the connected components which are claimed – in the theorems above – to be point components are proved to be so by using the method of divergence. In the cases where the method of divergence fails because the series are convergent the components are never point components as mentioned in the remarks above. These results are proved by using another technique, the theory of Mandelbrot-like families of quadratic-like mappings.

8. The Tableaux technique

In this section we deal with the point components in the dynamical planes. We start by the cubic case and comment on the technical difficulties in the quadratic case afterwards.

8.1. The cubic case.

In the cases where the modulus of an annulus of any depth can take only finitely many positive values there is nothing more to prove. The difficult cases are those where the moduli are not bounded away from 0. The tableaux technique was invented to deal with these cases.

The *tableau* $\mathcal{T}(x)$ of $x \in K_f$ is the double array of puzzle-pieces

$$T_{n,k}(x) = P_n(f^{\circ k}(x)) \quad \text{for} \quad n \geq 0, \; k \geq 0$$

with one column associated to each $f^{\circ k}(x)$ and one row associated to each depth in the puzzle. The polynomial f is mapping $T_{n,k}(x)$ onto $T_{n-1,k+1}(x)$. The tableau $\mathcal{T}(\omega_0)$ is called the *critical tableau*.

In order to estimate $\sum_n \mathrm{mod} A_n(x)$ it is sufficient to know the position of the critical pieces in the tableau. We shall therefore draw a marked grid to represent a given tableau. In the marked grid a *critical position* (n, k) is marked by a dot and a non-critical position is unmarked. Note that the 0th row is marked for any tableau and that the 0th column is marked for the critical tableau.

PROPOSITION 16 (TABLEAU RULES). *Let f be a cubic polynomial with the critical orbit of ω_0 bounded and the other critical orbit unbounded. The tableaux associated to f obey the following three rules:*

(T1) *The critical positions form unbroken vertical lines stretching out from the 0-row of the tableau; i.e. if $T_{n,k}(x) = C_n$ then $T_{j,k}(x) = C_j$ for $0 \leq j \leq n$.*
(T2) *If $T_{n,k}(x) = C_n$ for some x then $T_{i,k+j}(x) = P_{i,j}(\omega_0)$ for all $i + j \leq n$.*
(T3) *If for some $n \geq k > 0$ $T_{n-k+1,k}(\omega_0)$ is critical and $T_{n-j+1,j}(\omega_0)$ is not critical for $0 < j < k$ and if for any x and any m $T_{n,m}(x)$ is critical and $T_{n+1,m}(x)$ is not critical then $T_{n-k+1,k+m}(x)$ is not critical.*

The first rule is satisfied since the critical pieces are nested. The second rule is also immediate since it only reflects that a puzzle-piece which is the image under some iterates of f of a critical piece of some depth is the same wherever it occurs. The hypothesis in the third rule states that the mapping $f^{\circ k}$ from C_n to C_{n-k} has degree 2 and that the preimage by $(f|_{C_n})^{-k}$ of the piece C_{n-k+1} is C_{n+1}. Hence there cannot be a non-critical preimage as well.

It is important to notice that the marked grids really capture the essence of the dynamics as expressed in the following theorem.

THEOREM 17 (REALIZATION). *(a) Any critical marked grid satisfying the three rules above can be realized as the critical tableau for some cubic polynomial f.*
(b) Any marked grid satisfying the three rules above with respect to a given critical marked grid can be realized as the tableau $\mathcal{T}(x)$ for some $x \in K_f$ with f as in part (a).

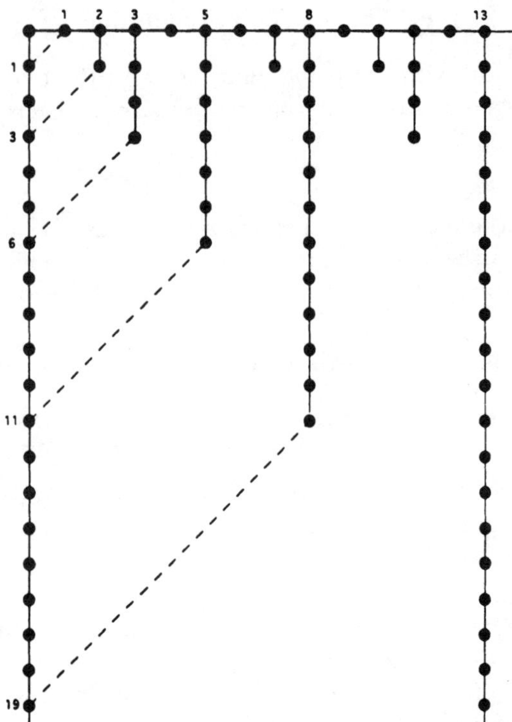

Figure 22. The critical grid of a cubic Fibonacci polynomial.

The critical tableau is said to be *periodic* if all positions in some column $k > 0$ are critical. The critical tableau for a polynomial f is periodic if and only if the critical component $K_f(\omega_0)$ is periodic.

Suppose the critical tableau is not periodic. We say that the critical tableau is *recurrent* if for any n there is a column $k(n)$ with critical position at depth n.

If f has a non-recurrent critical tableau then there exists an N so that the post-critical orbit $\{f^{on}(\omega_0)\}_{n>0}$ does not intersect C_N. One can show that all annuli $A_j(x)\,, j \geq 0,\ x \in K_f$, in this case have modulus bounded away from 0 and there is nothing more to prove.

If f has a recurrent critical tableau then the orbit of ω_0 intersects C_n for every n.

8.1.1. Cubic Fibonacci polynomials. As an example we shall consider a special critical marked grid which involves the Fibonacci numbers. This marked grid is therefore called the *Fibonacci grid* (see [**BH2**] Chapter 12) and any polynomial realizing the Fibonacci grid is called a *cubic Fibonacci polynomial.*

Let a_j denote the Fibonacci numbers:

$$a_0 = 1, \ a_1 = 2 \quad \text{and} \quad a_{j+1} = a_j + a_{j-1}.$$

The Fibonacci grid is constructed so that for any $j \geq 0$ the position

$$(a_{j+1} - 2, a_j) \text{ is critical and the position } (a_{j+1} - 1, a_j) \text{ is non-critical.}$$

Note that this information is sufficient to specify a unique critical grid satisfying the three tableau rules. (Compare with Figure 22.) Following the dotted line it is easy to see that $\mathrm{mod}A_{a_{j+1}-2}(\omega_0) = 1/2^{j+1} \ \mathrm{mod}A_0$.

Note that the Fibonacci polynomials have a non-periodic critical tableau which is recurrent.

8.1.2. Critically recurrent tableaux. For a critically recurrent tableau one defines the function $\tau : \mathbb{N} \to \mathbb{N} \cup \{0\}$ as follows:

$$\tau(n) = \max\{i \mid (i,j) \text{ is a critical position and } i + j = n, \ j > 0 \ \}.$$

This definition is due to Yoccoz. Moreover he formulates the tableaux rules in terms of rules of the τ-function.

The following lemmas can be proved from the three tableau rules.

LEMMA 18. *If* $\sum_n \mathrm{mod}A_n(\omega_0) = \infty$ *then* $\sum_n \mathrm{mod}A_n(x) = \infty$ *for all* $x \in K_f$.

LEMMA 19. *Consider a critically recurrent tableau. For any* $m \geq 0$ *there are at least two elements in* $\tau^{-1}(m)$.

LEMMA 20. *Suppose the critical tableau is periodic. Then* $\sum_n \mathrm{mod}A_n(\omega_0)$ *is convergent. Moreover* $\sum_n \mathrm{mod}A_n(x)$ *is convergent if and only if there exists a* $k \geq 0$ *such that* $T_{n,k}(x) = P_n(\omega_0)$ *for all* n.

The first lemma ensures that in order to prove the total disconnectedness of K_f it suffices to prove divergence of the series of moduli of annuli surrounding the critical point. We have already mentioned that this is only difficult in the critically recurrent case. Note that if $i \in \tau^{-k}(0)$ for some k then $\mathrm{mod}A_i(\omega_0) = 1/2^k \ \mathrm{mod}A_0$. The second lemma ensures that the number of elements in $\tau^{-k}(0)$ is at least 2^k. Set $I = \bigcup_{k \geq 0} \tau^{-k}(0)$. Then it follows that $\sum_{i \in I} \mathrm{mod}A_i(\omega_0) = \infty$. It is this balance between the decrease in the modulus and the minimum number of annuli realizing this value of the modulus which guarantees the divergence.

The third lemma ensures that for any cubic polynomial f with one bounded and one unbounded critical orbit all the connected components of K_f which are not eventually mapped onto the critical component $K_f(\omega_0)$ are point components. This completes the proof of Theorem 12 and part (b) of the result mentioned in the remark.

Note that for the critical Fibonacci grid we have exactly two elements in $\tau^{-1}(m)$ for each m. The Fibonacci grid was invented in order to show that this possibility can occur.

Yoccoz distinguishes between two different cases in the critically recurrent situation, defined as follows: a critically recurrent tableau is said to be *critically intermittent recurrent* if $\liminf_n \tau(n) < \infty$ and *critically persistent recurrent* if $\liminf_n \tau(n) = \infty$.

Note that the critical Fibonacci grid is critically persistent recurrent.

Remarks. (1) Theorem 12 can be extended to the case where f is a polynomial with unbounded critical orbits for all critical points but one. The tableau argument is exactly suited for a situation with one bounded critical orbit. It is essential to recall that we always count critical points with multiplicity, so that one bounded critical orbit means a bounded orbit of an ordinary critical point ω_0 for which $f''(\omega_0) \neq 0$.

(2) Most of the argumentation goes through in the case with unbounded critical orbits for all critical points but one of higher multiplicity. But the argumentation may fail in the critically persistent recurrent case. Suppose $m > 2$ and that the derivatives $f^{(j)}(\omega_0)$ are zero for $j = 1, 2, \ldots, m - 1$ and non-zero for $j = m$. If $i \in \tau^{-k}(0)$ for some k then $\mathrm{mod}A_i(\omega_0) = 1/m^k \, \mathrm{mod}A_0$. There is no longer a balance between the decrease in the modulus and the minimum number of annuli realizing this value of the modulus. For example the series $\sum_n \mathrm{mod}A_n(\omega_0)$ is convergent for $m > 2$ in the Fibonacci case. No conclusion can be drawn and it is in fact not known if the Julia set is a Cantor set in this situation.

8.2. The quadratic case.

In the quadratic case we work with tableaux as well. The 0th row is not critical, but contains critical positions. Recall that an annulus $A_n(x)$ is called critical if $P_{n+1}(x) = C_{n+1}$ and semi-critical if $P_n(x) = C_n$, but $P_{n+1}(x) \neq C_{n+1}$. We define the position (n, k) in the tableau $\mathcal{T}(x)$ to be critical if $T_{n+1,k}(x) = C_{n+1}$, semi-critical if $T_{n,k}(x) = C_n$ and $T_{n+1,k}(x) \neq C_{n+1}$ and non-critical otherwise.

The tableaux associated to a quadratic polynomial f_c with $c \in M^*_{r/s}$ obey the following five tableaux rules. The second is unchanged, the first and the third with obvious changes:

(T1*) A column which is neither entirely critical nor entirely non-critical has a unique semi-critical position so that every position above is critical and every position below is non-critical.

(T2) If $T_{n,k}(x) = C_n$ for some x then $T_{i,k+j}(x) = P_{i,j}(\omega_0)$ for all $i + j \leq n$.

(T3*) If for some $n \geq k > 0$ $T_{n-k,k}(\omega_0)$ is critical and $T_{n-j,j}(\omega_0)$ is non-critical for $0 < j < k$ and if for any x and any m $T_{n,m}(x)$ is semi-critical then $T_{n-k,k+m}(x)$ is semi-critical.

(T4*) There can be at most $s - 1$ consecutive non-critical positions in the 0th row.

(T5*) There are no semi-critical positions in the rows $1, 2, \ldots, s - 1$.

For a critical tableau the notions of periodic, recurrent and non-recurrent are unchanged. The analysis split again into three cases. The critical tableau is periodic if and only if the critical component $K^*_c(0)$ is periodic and in that case the polynomial f_c is renormalizable. If f_c has a non-recurrent critical tableau then one can prove that all the series $\sum \mathrm{mod}A_n(x)$ are divergent. If the critical tableau is recurrent one defines the function τ as before. The Lemmas 18 – 20 are unchnaged.

Most of the technical difficulties in the quadratic case come from the un-desirable annuli: the degenerate and the semi-critical. Suppose the critical tableau is recurrent. One can prove that there always exists a non-degenerate critical annulus. Suppose $\mathrm{mod}A_{n_0}(0) > 0$. Set

$$I = \bigcup_{k \geq 0} \tau^{-k}(n_0) \quad \text{and} \quad I^* = \{i \in I \mid (\tau(i), i - \tau(i)) \text{ is critical }\}.$$

Then one can prove that the series

$$\sum_{i \in I^*} \mathrm{mod}A_i(0)$$

is divergent. The argumentation is as follows: there is at least one element in $\tau^{-1}(m) \cap I^*$ for all $m \in I^*$. If there is exactly one element for a given m then there is at least two elements in $\tau^{-1}(n) \cap I^*$ for all $n \in \tau^{-\ell}(m), \ell > 0$. Note that if $i \in \tau^{-k}(n_0) \cap I^*$ for some k then $\mathrm{mod}A_i(0) = 1/2^k \ \mathrm{mod}A_{n_0}(0) > 0$. We see that even when we disregard all semi-critical annuli we still have, starting from a sufficiently deep level, the correct balance between the decrease in the modulus and the minimum number of annuli realizing the modulus.

8.2.1. Quadratic Fibonacci polynomials. Consider the modified critical grid obtained from the cubic critical Fibonacci grid in Figure 22 by erasing the 0th row. A quadratic polynomial in the limb $M_{1/2}$ realizing this modified critical Fibonacci grid is called a *quadratic Fibonacci polynomial*. This critical tableaux is recurrent and non-periodic. Hofbauer and Keller have (independently) defined a Fibonacci uni-modal map as a map where the closest returns of the critical point occur at the Fibonacci moments, see [**HK**]. Lyubich and Milnor have studied Fibonacci polynomials in more details and proved that the two definitions coincide when real quadratic polynomials are considered, see [**LM**].

Remark. Recently Nowicki and van Strien [**NS**] have announced that for each sufficiently large even integer d there exists $c \in \mathbb{R}$ such that the polynomial $z \mapsto z^d + c$ has *Fibonacci dynamics* and the Julia set is of positive Lebesgue measure. This provides the first examples of polynomials with Julia set of positive Lebesgue measure. The only other known examples among rational mappings are those where the Julia set is the whole Riemann sphere.

9. Final remarks

It should be evident from the previous description that there is an astonishing resemblance between the considered quadratic and cubic polynomials. The combinatorial resemblance between quadratic polynomials in $M_{1/2}$ and cubic polynomials in $B_{\rho,0}$ for some $\rho > 0$ was found by Yoccoz already in 1986. Note that if $f \in B_{\rho,0}$ then the postcritical set $\{f^{\circ n}(\omega)\}_{n>0}$ is contained in the ray $\mathcal{R}_f(0)$. As ρ tends to 0 the critical point ω escapes slower and slower to ∞. In the limit the critical point does not escape at all; instead the critical value $f(\omega)$ is fixed and moreover the landing point of the zero ray. In [**BD**] we proved (among other things) that the limb $M_{1/2}$ is homeomorphic to this set of cubic polynomials, i.e. cubic polynomials in the connectedness locus $\mathcal{C}$ for which one critical point

ω is pre-fixed and the critical value $f(\omega)$ is the landing point of one fixed ray. Intuitively one can imagine the motion in $\mathbb{C}^2$ of $B_{\rho,0}$ as ρ tends to 0. For each $\rho > 0$ the set $B_{\rho,0}$ consists of infinitely many connected components, namely countably many copies of M and uncountably many point components. In the limit all these components come together to form a connected set homeomorphic to $M_{1/2}$ on the boundary of $\mathcal{C}$.

In [**BF**] we are pursuing an analogous resemblance between quadratic polynomials and higher degree polynomials of the form $f_\lambda(z) = \lambda z(1 + \frac{z}{s})^s$. Let L_s denote the connectedness locus of this family and let $L_{s,0}$ denote the 0-limb of L_s, i.e. the polynomials with the multiple critical point $\omega = -s$ being pre-fixed and the critical value $f_\lambda(\omega)$ being the landing point of one fixed ray. We prove that for any $s \geq 2$ the limb $L_{s,0}$ is homeomorphic to any of the limbs $M_{r/s}$ of the Mandelbrot set. In particular limbs in the Mandelbrot set with the same denominator are homeomorphic.

References

[B] B. Branner, *The Mandelbrot set, Chaos and Fractals. The Mathematics Behind the Computer Graphics*, Proc. of Symposia in Appl. Math., AMS, 1989, pp. 75 – 105.

[BD] B. Branner and A. Douady, *Surgery on complex polynomials, Holomorphic Dynamics, Proc. Mexico, 1986*, Lecture Notes in Math., Springer, 1988, pp. 11 – 72.

[BF] B. Branner and N. Fagella, *Homeomorphisms of the limbs of the Mandelbrot set*, In preparation.

[**BH1**] B. Branner and J. H. Hubbard, *The iteration of cubic polynomials. Part I: The global topology of parameter space*, Acta Math. **160** (1988), 143 – 206.

[**BH2**] B. Branner and J. H. Hubbard, *The iteration of cubic polynomials. Part II: The patterns and parapattern*, Acta Math. **169** (1992), 229 – 325.

[D] A. Douady, *Descriptions of compact sets in* $\mathbb{C}$, *Topological Methods in Modern Mathematics*, Publish or Perrish, 1993, pp. 429 – 465.

[**DH1**] A. Douady and J, H, Hubbard, *Etude dynamique des polynômes complexes, I, II, Publication Orsay*, 1984, 1985.

[**DH2**] A. Douady and J. H. Hubbard, *On the dynamics of polynomial-like mappings*, Ann. Ecole Norm. Sup. (4) **18** (1985), 287 – 343.

[**HK**] F. Hofbauer and G. Keller, *Some remarks on recent results about S-unimodal maps*, Ann. Institut Poincaré **53** (1990), 413 – 425.

[H] John H. Hubbard, *Local connectivity of Julia sets and bifurcation loci: Three theorems of J.-C. Yoccoz, Topological Methods in Modern Mathematics*, Publish or Perrish, 1993, pp. 467 – 511.

[**HJ**] J. Hu and Y.Jiang, *The Julia set of the Feigenbaum quadratic polynomial is locally connected*, Preprint, CUNY (1993).

[J] Y.Jiang, *Infinitely Renormalizable Quadratic Julia Sets and Bifurcation Loci*, Preprint, CUNY (1993).

[K] L. Keen, *Julia Sets, This Proceedings*.

[L] M. Lyubich, *Geometry of Quadratic Polynomials: Moduli, Rigidity and Local Connectivity*, Preprint, IMS at Stony Brook (1993).

[LM] M. Lyubich and J. Milnor, *The Fibonacci Unimodal Map,*, Journal of the AMS **6** (1993), 425 – 457.

[M] J. Milnor, *Local connectivity of Julia Sets: Expository Lectures*, Preprint, IMS at Stony Brook (1992).

[NS] T. Nowicki and S. van Strien, *Polynomial maps with a Julia set of positive Lebesgue measure: Fibonacci maps*, Preprint, Amsterdam (1994).

[P] C. Lunde Petersen, *Local connectivity of some Julia sets containing a circle with an irrational rotation*, Preprint, I.H.E.S. (1994).

[S] D. Sørensen, *Complex Dynamical Systems: Rays and Non-Local Connectivity*, Ph.D. Thesis, Tech. University of Denmark (1994).

[T]   Tan Lei, *Voisinages connexes des points de Misiurewicz*, Ann. de l'Inst. Fourier **42** (1992), 707 – 735.

[Y]   J.-C. Yoccoz, *Local connectivity of the Mandelbrot set*, In preparation.

MATHEMATICAL INSTITUTE, BUILDING 303, THE TECHNICAL UNIVERSITY OF DENMARK, DK–2800 LYNGBY, DENMARK

E-mail address: branner@mat.dtu.dk

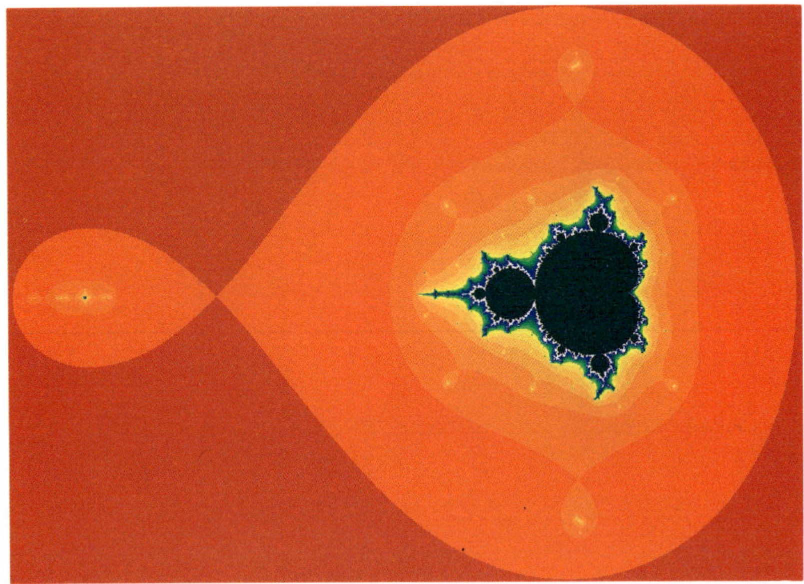

COLOR PLATE 1. This shows the parameter space Λ_0 for cubic polynomials. The annuli $BA_n(\lambda)$ are seen in colors; different colors correspond to different levels. The largest connected components of B_0, homeomorphic to copies of the Mandelbrot set, are seen in black.

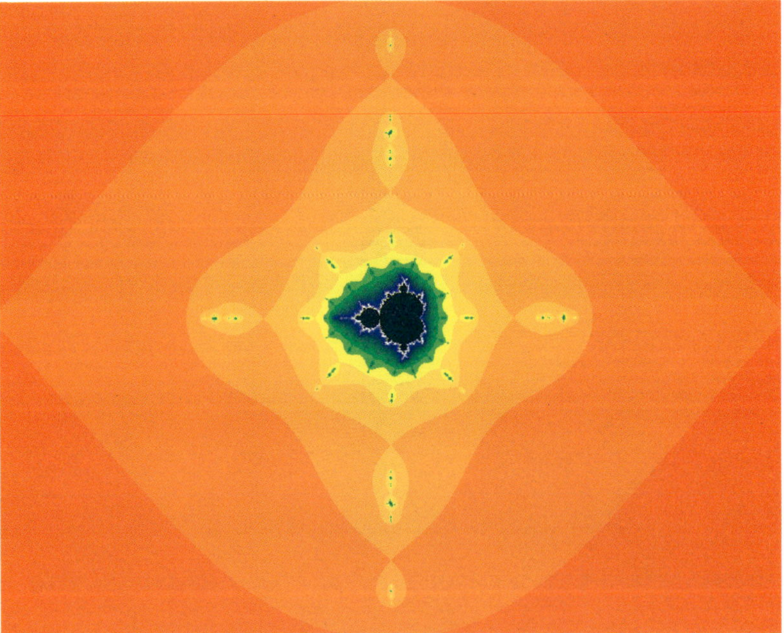

COLOR PLATE 2. This is a blow up of Color Plate 1 around the second largest connected component of B_0.

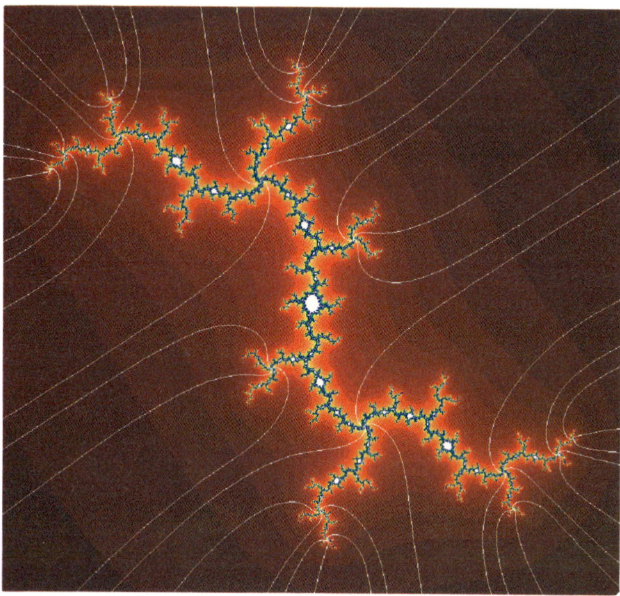

COLOR PLATE 3. This shows the dynamical plane for the quadratic polynomial f_c with $c = 0.0150745 + i0.848364$. This c-value belongs to MV_3' in $M_{1/3}$. The filled Julia set K_c is shown together with the external rays defining the puzzle of depth 4.

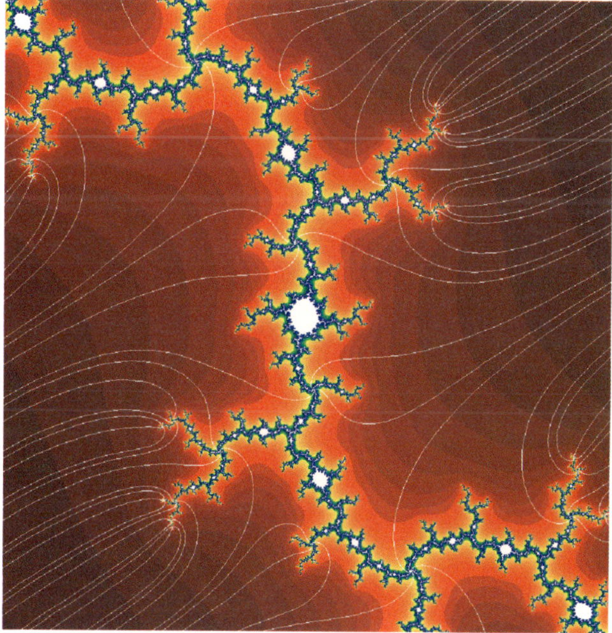

COLOR PLATE 4. This is a blow up of Color Plate 3 around the critical point (from α_c in the upper left corner to $-\alpha_c$ in the lower right corner) together with the external rays defining the puzzle of depth 7.

Proceedings of Symposia in Applied Mathematics
Volume **49**, 1994

Julia Sets of Rational Maps

LINDA KEEN

1. Preamble

This is the second time the American Mathematics Society has offered a *Short Course* in holomorphic dynamics; the first was at the Centennial meeting in the summer of 1988. At that time the literature for those interested in learning the subject was quite thin. There are now several good sources for learning this material including [**2, 8, 9, 13, 14**]. In this article we again give a brief survey of the theory of Julia sets for rational maps. Many of the open problems presented in the first short course on this subject in 1988 (see [**9**] have now been solved. Although we have space here only to sketch proofs, we shall slightly modify our approach in [**9**] to better emphasize the ideas used in the new work. We have tried to include the current literature in the bibliography. Fuller bibliographies may be found in [**2, 8, 14**].

2. Introduction

A *rational map* is a finite to one complex analytic map of the Riemann sphere $\hat{\mathbb{C}}$ to itself. It can be written as a quotient of polynomials in lowest terms:

$$ R(z) = P(z)/Q(z) = \frac{a_0 z^d + \cdots + a_d}{b_0 z^{d'} + \cdots b_{d'}}, \quad Q(z) \not\equiv 0. $$

The quotient is said to be in lowest terms if P and Q have no common factors; a lowest terms representation is unique up to multiplication of the numerator and denomimator by a nonzero constant. The *degree* of R can be defined either as the maximum (and typical) number of preimages of a given point z or as the maximum of the degrees of P and Q. In this article, we shall only consider maps whose degree is greater than 1. The degree 1 case is particularly simple and easily understood (see e.g.[**2**], Chap. 1).

We shall be interested in the holomorphic dynamical systems that arise by iterating such maps. This means we want to describe the behavior of the iterates

---

1991 *Mathematics Subject Classification.* Primary 30D05,58F23.
Research partially supported by NSF grant DMS-8902881

$$R^{\circ n}(z) = \underbrace{R \circ R \circ \cdots \circ R(z)}_{n \; times}$$

at points of $\hat{\mathcal{O}}$; that is, we want to describe the structure of the orbits

$$\{z_0, z_1 = R(z_0), z_2 = R^{\circ 2}(z_0), \dots\}$$

for various points z_0. As Devaney discussed in his article, if the map is contracting at z_0, orbits of points near z_0 travel together predictably; if, however, the map is expanding at z_0, nearby points may cease to be close to the orbit of z_0 and we see chaotic behavior.

Contraction and expansion may be measured in terms of the derivative of the iterated map: it contracts if its derivative has modulus less than one and expands if it has modulus greater than one. For a rational map of degree $d \geq 2$, the derivative is expanding on average since if we normalize $\hat{\mathcal{O}}$ to have total area 1 with respect to the spherical metric and consider the integral:

$$I = \int_{\hat{\mathbf{C}}} \|(R^{\circ n})'\|^2 \sim d^n,$$

we see that I tends to infinity with n. The map R cannot expand everywhere however since the derivative R', (and by the chain rule the derivative of any iterate) certainly vanishes at some points. These points are called *critical points* and, as we shall see, near a critical point the behavior of R is very much like the behavior near the origin of $z \mapsto z^2$. The Julia set is the boundary between the competing contracting forces. It is the contradictory effect of these forces that makes the dynamics so interesting.

3. A simple example

Let us begin with a very simple example.

EXAMPLE. $P(z) = z^2$.

First, let z_0 be any point outside the unit circle, $|z_0| > 1$. Since applying $P(z)$ squares the modulus and doubles the argument of z_0 we see that $|z_n| = |z_0|^{2^n} \to \infty$; in fact, the orbit of every point outside the unit circle converges toward infinity.

Next, let z_0 be any point inside the unit circle, $|z_0| < 1$. Again $|z_n| = |z_0|^{2^n}$, but now $|z_n| \to 0$ and all such orbits converge toward the origin.

Finally, let z_0 be on the unit circle, $z_0 = e^{2\pi i \alpha}$. Now the orbit structure depends on α. Let's consider some examples.

If $\alpha = 0$, z_0 is fixed by $P(z)$; if $\alpha = \pi$,

$$-1 \xrightarrow{P} 1 \xrightarrow{P} 1,$$

and the orbit consists of two points. Similarly, if $\alpha = p/2^k$, for any positive integers p and k, the orbit consists of the $k+1$ points

$$e^{2\pi i p/2^k} \xrightarrow{P} e^{2\pi i p/2^{k-1}} \xrightarrow{P} e^{2\pi i p/2^{k-2}} \xrightarrow{P} \ldots \xrightarrow{P} -1 \xrightarrow{P} 1.$$

If $\alpha = p/q$ with q odd, the orbit is periodic. For example, if $p/q = 2/7$, the orbit is

$$e^{2\pi i/7} \xrightarrow{P} e^{4\pi i/7} \xrightarrow{P} e^{8\pi i/7} \xrightarrow{P} e^{2\pi i/7}.$$

When q is even the orbit is preperiodic. For example if $p/q = 1/12$ the orbit is

$$e^{\pi i/6} \xrightarrow{P} e^{\pi i/3} \xrightarrow{P} e^{2\pi i/3} \xrightarrow{P} e^{4\pi i/3} \xrightarrow{P} e^{2\pi i/3}.$$

When α is irrational the orbit is infinite; $z_n \neq z_m$ for any $n \neq m$, but $|z_n - z_m|$ is very close to zero infinitely often.

In this simple example of the squaring map, we see various different types of orbits; periodic, preperiodic, infinite and converging to a fixed point, and infinite and chaotic. As we shall see below, these are essentially the only kinds of orbits that occur.

4. Definitions and Background

4.1. Rational maps. In this section we define the basic concepts involving rational maps:

DEFINITION. *R is a local homeomorphism at all but finitely many points; those points where it fails to be a homeomorphism are* critical points. *In a neighborhood of a critical point, R is a k to 1 ramified covering of its image, $k \leq d$. The critical points in the complex plane are the zeros of the derivative R'. To see if ∞ is a critical point we make an appropriate change of variable.*

DEFINITION. *A* critical value *is the image of a critical point.*

A point has the maximum number of distinct preimages if and only if it is not a critical value. Critical values have fewer preimages and it can be shown by the Riemann-Hurwitz theorem [**3**] that counted properly, a degree d rational map has exactly $2d - 2$ critical values.

DEFINITION. *The* post-critical set *is the closure of the set of orbits of the critical points.*

In the analysis of the dynamical system defined by iterating the map R, the role of the post-critical set is crucial. Away from it, the iterates are covering maps, and the covering properties are very important.

DEFINITION. *A* family of complex analytic functions *$\{f_\alpha\}$, defined on a domain D is called a* normal family *if every infinite subset contains a subsequence which converges uniformly on every compact subset of D.*

REMARK. *The limit function is generally not a member of the family. In the example above, where the members of the family are the iterates of z^2 and D is either the interior or the exterior of the unit circle, the limit functions are the constants 0 and ∞, and hence do not belong to the family.*

An important characterization of normal families is given by Montel's theorem:

THEOREM 4.1 (MONTEL). *Let F be a family of analytic functions defined on a domain D. If the union $\cup_{f \in F} f(D)$ omits three points in $\hat{\mathbb{C}}$, Then F is a normal family.*

Using these ideas we may define what we mean by the predictable or chaotic behavior of an orbit.

DEFINITION. z *is a* stable *point for R if there is a neighborhood U of z such that $\{R^{\circ n}\}$ form a normal family on U. Let Ω_R (or Ω if there is no confusion) denote the stable set of R.*

As we shall see, the stable points have predictable behavior. Moreover, it is clear from the definition that nearby points have similar behavior. In the literature, the stable set is sometimes called the *normal set* or the *Fatou set*.

DEFINITION. *The* unstable set *is the complement of Ω_R; it is also known as the* Julia set *and is denoted by J_R (or simply J).*

As we shall see, points in the Julia set exhibit chaotic behavior in the sense described in Devaney's article in this collection.

The set Ω_R is open by definition; it must either be empty or dense in $\hat{\mathbb{C}}$ and it is often not connected. The Julia set is therefore closed and we will see in proposition 4.3 that it is not empty.

Using the definition, we easily prove that Ω_R and J_R are both *completely invariant* under R; that is:

PROPOSITION 4.2. *If z is in Ω_R, then its image and and all its preimages are also in Ω_R.*

PROOF. Let U be a neighborhood of z on which the iterates $R^{\circ n}$ form a normal family. By the open mapping theorem $R(U)$ is open. If $R^{\circ n_j}$ is a convergent sequence on U, $R^{\circ n_j - 1}$ is a convergent sequence on $R(U)$. Similarly, $R^{-1}(U)$ is open and on each component of $R^{-1}(U)$, $R^{\circ n_j + 1}$ is a convergent sequence. ∎

It now follows that J_R is also completely invariant.

PROPOSITION 4.3. *J_R is not empty.*

PROOF. Suppose it were. Then, for any z, we could find a neighborhood U and a subsequence of the iterates, $R^{\circ n_j}$, that converge to an analytic function $f : U \to \hat{\mathbb{C}}$, which by the monodromy theorem, can be continued analytically to $\hat{\mathbb{C}}$. It must therefore be a rational function of some finite degree. But, since f is the limit of functions whose degrees tend to ∞, it must have infinite degree and we have a contradiction. ∎

In the example where $R(z) = z^2$, Ω_R consists of the interior and the exterior of the unit circle and J_R is the circle itself.

An example in which $\Omega = \emptyset$ and $J = \hat{\mathbb{C}}$ is the function $R(z) = (z-2)^2/z^2$ first considered by Guckenheimer. This will follow from the classification theorem for stable domains.

Note that $R^{\circ k}$ is a rational map of degree d^k. It is clear from proposition 4.2 that $\Omega_{R^{\circ k}} = \Omega_R$ and therefore that $J_{R^{\circ k}} = J_R$.

If $z \in J$, there can be no neighborhood U for which the iterates of R form a normal family. Hence by Montel's theorem, $\cup_n R^{\circ n}(U) \supset \hat{\mathbb{C}} - E_R$, where E_R is a set containing at most 2 points; E_R is called the *exceptional set*, and is completely invariant.

It is fairly obvious that if we conjugate the squaring map by an affine map, the Julia set of the conjugated map is again a circle and the stable set consists of the interior and exterior of this circle. It is not quite as obvious, but it is not too difficult to show, that if a rational map is conjugated by a Moebius transformation (a degree 1 rational map), the qualitative structure of the stable and unstable sets is unchanged. A proof of this fact for quadratic polynomials can be found in Devaney's article.

Therefore, if $E_R = \{e\}$, we may conjugate so that $e = \infty$. Since E_R is completely invariant, e has only itself as preimage so it is critical with multiplicity $d - 1$, where d is the degree of R; hence R is a polynomial. It is easy to check that ∞ is always stable for a polynomial since for $|z| > M$, M large, $|R^{\circ n}(z)| \sim |z|^{d^n} \to \infty$. If $E_R = \{a, b\}$, we may conjugate so that $a = 0, b = \infty$. Again since E_R is completely invariant both points are critical with multiplicity $d - 1$, and $R = cz^{\pm d}$ for some $c \neq 0 \in \mathbb{C}$. Both points of E are clearly stable. Any map not conjugate to one of these maps has an empty exceptional set.

It is now easy to prove:

PROPOSITION 4.4. *If J contains a non-empty open subset then $J = \hat{\mathbb{C}}$.*

PROOF. Let U be an open set in the interior of J. By Montel's theorem, $\cup_n R^{\circ n}(U) \supset \hat{\mathbb{C}} - E$ and by the invariance of J, $J \supset \hat{\mathbb{C}} - E$. Since J is closed and E contains at most 2 points, $J = \hat{\mathbb{C}}$. ∎

We are now able to prove that the action of a rational map restricted to its Julia set is chaotic.

PROPOSITION 4.5. *If $w \in J$, then $\cup_n R^{\circ -n}(w)$ is dense in J.*

PROOF. Let U be a neighborhood of some $z \in J$. There is some k so that $w \in R^{\circ k}(U)$; therefore, for some branch of $R^{\circ -k}$, $R^{\circ -k}(w) \in U$. ∎

REMARK. *This proof actually shows more. It shows that if z is not in the exceptional set E, then the inverse images, $R^{\circ -n}(z)$ accumulate onto the Julia set. It is this property of the Julia set which is used most often in creating the computer pictures which approximate the Julia sets.*

PROPOSITION 4.6. *J is an infinite set (hence by the above, it is perfect).*

The proposition follows from the following lemma:

LEMMA 4.7. *If S is a non-empty completely invariant set under the rational map R, then it contains one, two or infinitely many points.*

PROOF. Suppose S contains finitely many points, $S = \{a_0, a_1, \ldots, a_k\}$. Since S is completely invariant, there is some iterate $R^{\circ p}$ that fixes each point in S. To simplify notation, let d be the degree of $R^{\circ p}$. Now by complete invariance, each point of this set has only itself as preimage under $R^{\circ p}$ and therefore each of these points is critical with multiplicity $d - 1$. Since the total number of critical points counted with multiplicities is $2d - 2$, k is at most 2. ∎

PROOF. [of proposition 4.6] Recall our discussion of exceptional sets above. A completely invariant set containing only one or two points must be the exceptional set of a polynomial. Since points in the exceptional set are always stable the non-empty completely invariant set J must contain infinitely many points. ∎

PROPOSITION 4.8. *J does not properly contain any closed completely invariant set.*

PROOF. Let $K \subset J$ be closed and completely invariant. For any $w \in K$, the inverse orbit of w is contained in K by invariance and is dense in J by proposition 4.5. Since K is closed $K = J$. ∎

5. Periodic Points

A *period-p periodic cycle* is a sequence of distinct points, $\{z_0, z_1, \ldots, z_{p-1}\}$ such that $z_i = R(z_{i-1})$ and $z_0 = R(z_{p-1})$. Each point in the cycle is a fixed point of the *first return map* $f = R^{\circ p}$. From the chain rule, it follows that the same value is obtained by evaluating the derivative $f' = (R^{\circ p})'$ at each of the z_i's; this common value, λ , is the *eigenvalue* of the cycle.

DEFINITION. *A periodic point or cycle is called:*

$$
\begin{array}{ll}
\text{attracting } \textit{if} & 0 < |\lambda| < 1 \\
\text{super-attracting } \textit{if} & |\lambda| = 0 \\
\text{repelling } \textit{if} & |\lambda| > 1 \\
\text{neutral } \textit{if} & |\lambda| = 1.
\end{array}
$$

For the squaring map, the only stable periodic points are the super-attracting fixed points 0 and ∞. All the other periodic points are repelling and unstable and form a dense subset of the unit circle. This phenomenon is characteristic for Julia sets of rational maps. We have

THEOREM 5.1. *The Julia set of a rational map R is the closure of the repelling periodic points.*

As we shall see, the maps with the property that the Julia set is disjoint from the post-critical set form an important class, the hyperbolic maps. We give the proof for these maps.

PROOF. Let $w \in J$, and choose a neighborhood U of w, disjoint from the post-critical set. There is an integer k, such that $R^{\circ k}(U) \supset U$, and since there are no critical points of $R^{\circ k}$ in U, there is a well-defined branch of $R^{\circ -k}$ such that $V = R^{\circ -k}(U) \subset U$. Since, by the Schwarz lemma, $R^{\circ -k}$ is non-expanding in the Poincaré metric, the sets $R^{\circ -nk}(U), n > 0$, form a nested sequence and thus their intersection contains an attracting fixed point for $R^{\circ -k}$. This is the required repelling periodic point for R. ∎

6. Classification of stable domains

As we saw, the behavior at the stable periodic points of the squaring map is easily understood. To appropriately generalize, we shall to study the local behavior in a neighborhood of a periodic point for an arbitrary rational map.

In what follows we assume the periodic point is at 0 and we write the first return map as a power series:

$$f(z) = \lambda z + a_k z^k + \ldots$$

with $k \geq 2$.

If $\lambda \neq 0$ and $|z|$ is small then the map "looks like" its linear part, $g(z) = \lambda z$; if $\lambda = 0$ it "looks like" the power map $g(z) = z^k$. We can make the concept "looks like" precise as follows.

DEFINITION. *The maps $f : U \to U$ and $g : V \to V$ are called* conformally conjugate *if there exists a homeomorphism $\phi : U \to V$ such that $\phi \circ f = g \circ \phi$.*

Note that iteration, and hence periodicity, are preserved under conjugation.

If f is conformally conjugate to its linear part in a neighborhood of a fixed point it is said to be *linearizable* there.

The idea of using conformal conjugation to study iteration goes back to Schröder. The point is that we can use the conjugation to give us a local model for the dynamical behavior. The following theorem enables us to fully understand the dynamics in a neighborhood of a stable periodic point.

THEOREM 6.1 (SCHRÖDER). *A periodic point of a rational map with multiplier $\lambda \neq 0$ is stable if and only if it linearizable.*

PROOF. [Sketch] Use the fact that at a stable point the iterates are uniformly equicontinuous. ∎

6.1. Attracting cycles. Now suppose z_0 is an an attracting period-p point of R.

THEOREM 6.2 (KOENIGS). [10] *If z_0 is an attracting period-p point of a rational map R then the first return map is linearizable at z_0. Moreover, the conjugating map is unique up to scaling.*

PROOF. [Sketch] With notation as above, set $\phi_n(z) = \lambda^{-n} f^{\circ n}(z)$; we have

$$\phi_n \circ f = \lambda \phi_{n+1},$$

so that if ϕ_n converges uniformly to some ϕ on U, ϕ is the required conjugation. The hypothesis $|\lambda| < 1$ can now be used to obtain the convergence. Uniqueness follows from the Schwarz lemma. ∎

REMARK. *If $|\lambda| > 1$ we may apply the above argument to the branch of f^{-1} fixing z_0. Since f maps the neighborhood U outside itself the conjugation is not well defined under iteration and doesn't model the dynamics.*

From theorem 6.2 we see that for each point z_k, $k = 0, \ldots p-1$ of an attracting periodic cycle, there is a maximal open stable domain I_k containing z_k. The set $I = \cup_0^{p-1} I_k$ is called the *immediate attracting basin* of the cycle. The set $A = \cup_{n>0} R^{\circ -n}(I)$ is called the *attracting basin*, and the cycle of domains is called an *attracting cycle*.

PROPOSITION 6.3. *The immediate attractive basin of an attractive cycle always contains a critical point that is attracted to the periodic points.*

PROOF. Let $I_k, k = 0, \ldots, p - 1$ be the p components of the immediate attractive basin so that $R(I_k) = I_{k+1 \bmod p}$. If there is no critical point in any I_k, all these maps are covering maps and hence must be isometries in the Poincaré metric. This is a contradiction since the cycle is attracting. ∎

Figure 1 shows the Julia set for an attracting cycle.

6.2. Super-attracting cycles. Suppose now that $z_0, z_1, \ldots z_{p-1}$ is a super-attractive cycle so that the eigenvalue $\lambda = 0$. With the notation above we have

THEOREM 6.4 (BÖTTCHER). [4] *If the fixed point z_0 of the first return map f is super-attracting, and if the first $k - 1$ derivatives vanish at z_0, there is a conformal conjugation ϕ defined on a neighborhood U of z_0 such that $\phi \circ f = (\phi(z))^k$. The conjugation ϕ is unique up to a multiplication by a $k - 1st$ root of unity.*

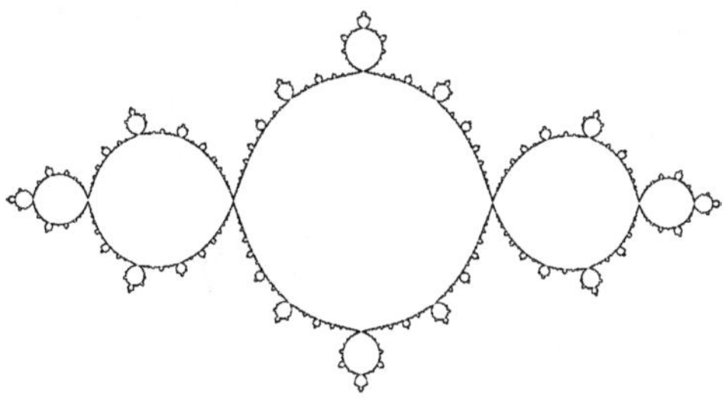

FIGURE 1. An attracting cycle

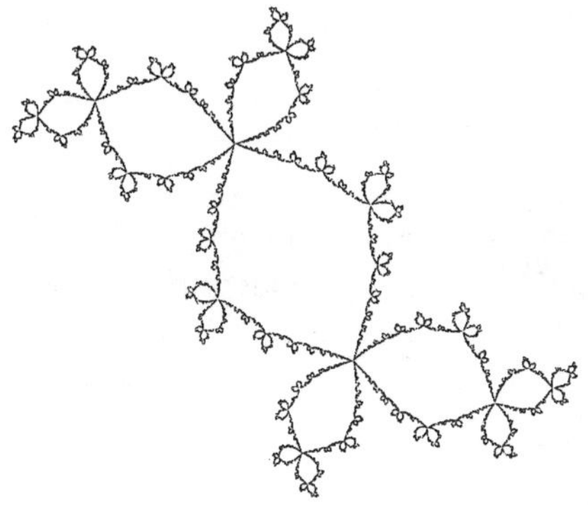

FIGURE 2. A super-attracting cycle

PROOF. [Sketch] Set $\phi_n(z) = f^{\circ n}(z)^{k^{-n}}$ so that ϕ_n satisfies

$$\phi_{n-1} \circ f = (f^{\circ n-1} \circ f)^{k^{-n+1}} = \phi_n'$$

and show that ϕ_n converges to the desired conjugation. ∎

Using the conjugacy it is easy to see that super-attractive points are stable. The attractive and immediate attractive basins are defined as they were for attractive cycles and in this case we call the cycle of periodic domains *super-attractive*. By definition there is a critical point in one of the domains of the immediate attractive basin.

Figure 2 shows a the Julia set for a super-attracting cycle.

6.3. Neutral periodic points.

6.3.1. *The rational case.* Suppose that z_0 is a neutral periodic point whose eigenvalue λ is an n-th root of unity. Such a point is called *parabolic*.

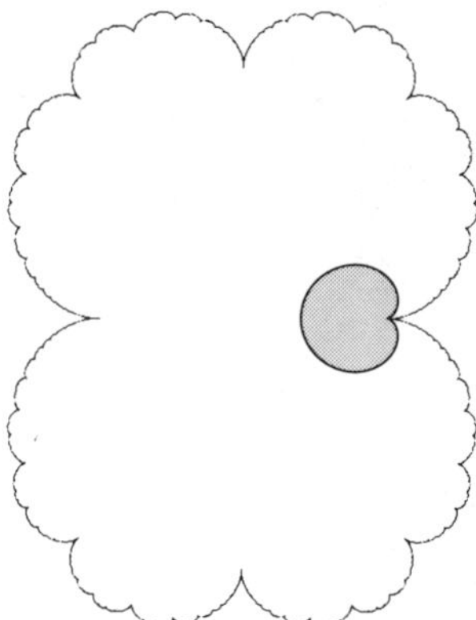

FIGURE 3. An attracting petal

To find the local behavior, we may again assume z_0 is a fixed point of the first
return map f. If z_0 were stable, by Schröder's theorem f^{on} would be conjugate
to the identity map in a neighborhood of z_0. Using power series expansions it
is easy to check that the only map conjugate to the identity is itself and so by
analytic continuation, f^{on} would be the identity everywhere, which is nonsense.

Therefore, we cannot model the dynamical behavior in a full neighborhood of
a parabolic point by the linear part of the rational map. We can, however, do
so in partial neighborhoods.

PROOF. [Sketch] Assume that $z_0 = 0$ and that $\lambda^n = 1$; the first return map
has the form:

$$f(z) = \lambda z + a_k z^k + \dots .$$

(i) First assume $\lambda = 1, k = 2$. We may conjugate f by scaling and inversion
     using a Moebius transformation h to obtain the following expansion near
     ∞ for the conjugated map:

$$g(z) = z + 1 + b/z + \dots .$$

(ii) Now prove that in some right half plane, $\Re z > K$, the map g is conjugate
      by a map ϕ to the translation $z \mapsto z + 1$. Next show that the domain
      of definition of ϕ can be extended to a domain Π bounded on the left
      by the parabola $y = 4K(K - x)$; we have $g(\Pi) \subset \Pi$. Therefore, if
      $P = h^{-1}(\Pi)$, $f(\bar{P}) \subset P \cup \{0\}$, and P is a forward invariant domain,
      symmetric about the real axis, with the fixed point 0 on its boundary.
      Moreover, $\cap_{n \geq 0} f^{on}(\bar{P}) = \{0\}$. The domain P is called an *attracting
      petal* (see Figure 3).
      Note that along the ray of the real axis emanating from the fixed point
      that is not contained in the petal, points are repelled. If the same ideas
      are applied to the inverse map we obtain a *repelling petal.* The attracting

FIGURE 4. The attracting petals of the Leau flower with $k = 3$

petal is contained in the stable set. The repelling petal contains both stable and unstable points.

(iii) Next assume $\lambda = 1$ but $k > 2$. In this case we make a change of variable, $z = \zeta^{1/k}$. We obtain a sequence of $2(k-1)$ alternately attracting and repelling petals. See Figure 4. The resulting picture is called the *Leau-Fatou Flower*.

(iv) Finally, if $\lambda = e^{2\pi i p/q}$, we apply the previous analysis to each of the q fixed points of $f^{\circ q}$; the attracting petals are not forward invariant but points in a given attracting petal cycle around through the attracting petals at each of the periodic points.

There are various proofs for this picture. One involves a careful estimate of the coefficients of the Taylor series g. Another method is an ingenious application of quasiconformal mappings developed by Yoccoz in his proof of theorem 6.6. The details of either proof are beyond the scope of this article but may be found in **[2, 8, 14]**. ∎

The *immediate attractive basin* of a parabolic cycle is defined as maximal set of domains attracted to the parabolic cycle whose boundaries contain points of the cycle. It contains the attractive petals. These periodic components are called a *parabolic cycle* of domains. The following is proved using the Schwarz lemma just as in the case of attracting cycles.

PROPOSITION 6.5. *The immediate attractive basin of a paraboic cycle always contains a critical point that is attracted to the periodic points.*

Figure 5 shows the Julia set for a map, named the *fat rabbit* by Douady, that has a parabolic cycle. The critical point is in the center of the central circular region. It is attracted to the fixed point which is on its boundary at the upper point where the three regions meet. It is difficult to completely show this part of

FIGURE 5. A parabolic cycle

the Julia set in a computer drawing since the map is repelling in the directions of the Julia set.

6.4. The irrational case. Suppose now that z_0 is a neutral periodic point with eigenvalue $\lambda = e^{2\pi i \alpha}$ where α is irrational. A long outstanding question was to characterize those values of α for which the point is linearizable. Sufficient conditions had been given first by Siegel and later by Brjuno, and recently, Yoccoz proved that for quadratic polynomials Brjuno's condition is also necessary [20]. The condition is:

THEOREM 6.6 (BRJUNO-YOCCOZ). *A sufficient condition for a rational map with a neutral periodic point with multiplier $\lambda = e^{2\pi i \alpha}$, α irrational, to be linearizable in a neighborhood of the point is*

$$\sum_1^\infty \frac{\log q_{n+1}}{q_n} < \infty$$

where p_n/q_n are the continued fraction approximants to α. If R is a quadratic polynomial the condition is also necessary.

The proof uses techniques from quasiconformal mappings and is beyond the scope of this article. It is still an important open question to find necessary conditions for arbitrary rational maps.

6.4.1. *Siegel disks.* When z_0 is a linearizable neutral periodic point for the first return map f, there is a conjugation $\phi\colon U \to \Delta$ that takes a neighborhood U of z_0 to the unit disk Δ and satisfies

$$\phi \circ f = e^{2\pi i \alpha} z.$$

The neighborhood U is foliated by forward invariant leaves that are the preimages of the circles $|\zeta| = r$ in Δ. The maximal domain for which the conjugation can be defined is called *a Siegel disk*. The neutral periodic cycle thus defines a periodic

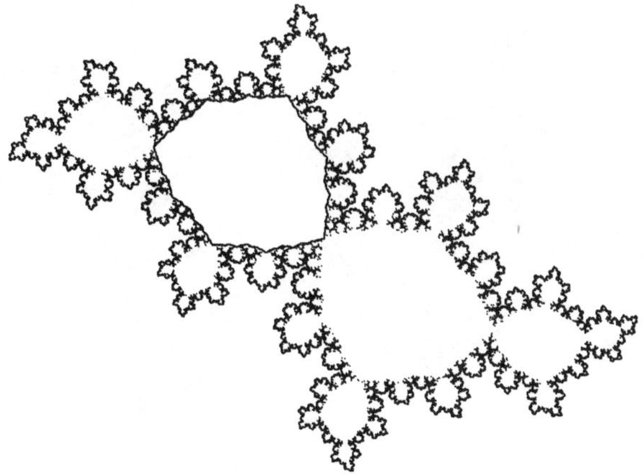

FIGURE 6. A Siegel Disk

cycle of Siegel disks. Any critical point that falls into one of these disks is trapped into a leaf of the foliation — but it is not necessarily true that a critical point does fall into one. Although he only suspected that these domains existed, Fatou proved

PROPOSITION 6.7. *If a rational map has a cycle of Siegel disks, then the forward orbit of some critical point must accumulate in its boundary.*

The proof uses the fact that away from the postcritical set the map is a regular covering and that there must be a forward invariant neighborhood U of the fixed point in the Siegel disk. If the proposition were false, we could find branches of the inverse along the boundary of the disk whose forward orbit would miss U which is impossible since the boundary is unstable. Figure 6 shows how the critical orbit accumulates on the boundary.

A critical point whose orbit accumulates in the boundary of a Siegel disk must be in the Julia set to begin with. Herman showed that there are Siegel disks whose boundary actually contains a critical point and those whose boundary does not. See [8] for a full discussion.

6.4.2. *Cremer points.* H. Cremer gave the first examples of neutral periodic points that are not linearizable so his name is now attached to any such. If U is a neighborhood of a Cremer point, the forward orbit of some critical point must visit U infinitely often. This is proved in [7]. This causes the Julia set to be very complicated in U. Douady and Hubbard have found examples of quadratic maps whose Julia sets are not locally connected in the neighborhood of a Cremer point.

6.4.3. *Herman rings.* The last type of periodic stable component can only occur for rational maps that are not polynomials. These components are called

FIGURE 7. A Herman ring

Herman Rings after Michel Herman who found the first examples. A Herman ring is a component of the stable set that is topologically an annulus and on which the first return map is conformally conjugate to an irrational rotation of a regular annulus. The images of the circles from the regular annulus form a forward invariant foliation in the Herman ring. It is clear that the ring can contain no periodic point. The proof that the boundary of a Herman ring $\mathcal{A}$ is contained in the accumulation set of the post-critical set is the same as it is for Siegel disks. The bounded component $\hat{\mathcal{O}} - \mathcal{A}$ is bounded by Julia set and so must contain a pole of the function; polynomials, therefore never have Herman rings.

Herman's examples belong to the family:

$$R(z) = e^{2\pi i\alpha} z^2 (z - a/1 - az)$$

for suitable choices of $0 < \alpha < 1$ and $a > 1$.

6.5. Wandering domains. The stable set Ω_R is divided into a collection of mutually disjoint maximal connected components by the Julia set. It is easy to see that if D_0 is one such component then $R(D_0)$ is also. One of the following must hold for each component D_0:

(i) the forward iterates $R^{\circ j}(D_0)$ are disjoint, in which case D_0 is called *wandering*, or

(ii) there exist integers $m, n \geq 0$, with $R^{\circ n}(D_0) = R^{\circ m}(D_0)$, and D_0 is called *eventually periodic.*

The qualitative and quantitative classification of eventually periodic domains was begun by Fatou and Julia. They studied the attracting, super-attracting and parabolic domains. Their analysis showed that the only other type of stable behavior that might exist would be the rotation domains (Siegel disks and Herman rings), although they couldn't prove such domains did actually exist. This was done by Siegel and Herman respectively. Fatou conjectured that wandering domains did not exist and therefore that the only stable behavior is characterized by the five models we described above. The classification was completed Sullivan and Shishikura. References and a full discussion may be found in [**2, 8, 14**].

In the early 1980's, Sullivan, [**17**], saw that the non-existence of wandering domains for rational maps was analogous to the fundamental finiteness theorem of Ahlfors for Kleinian groups and he gave a proof for both theorems using quasiconformal mappings.

THEOREM 6.8 (SULLIVAN). *Every component of the stable set of a rational map is eventually periodic.*

In view of theorem 6.8, there is a natural partition of the set of components of the stable set into equivalence classes given by $U \sim V$ if U and V eventually fall into the same periodic cycle. It follows easily from our discussion above that since each of the five models of eventually periodic stable behavior "uses" a critical point, and that since there are only finitely many critical points, there can be only finitely many equivalence classes of stable components. A more sophisticated result is the sharp upper bound given by Shishikura:

THEOREM 6.9 (SHISHIKURA). *A degree d rational map has at most $2d - 2$ cycles of periodic stable domains.*

It is no coincidence that this bound is precisely the number of critical points, counted with multiplicity. The proof is beyond the scope of this article.

As a consequence of the classification of stable domains we have:

COROLLARY 6.10. *If the post-critical set is finite and no critical point is periodic, then $J = \hat{\mathbb{C}}$.*

PROOF. In each type of stable domain except a super-attracting one the critical point must have an infinite forward orbit. ∎

Recall Guckenheimer's example: $R(z) = (z - 2)^2/z^2$. The critical points are at $z = 2$ and $z = \infty$ and we compute:

$$2 \to 0 \to \infty \to 1 \to 1.$$

7. Hyperbolic Rational Maps

An important class of rational maps are the *hyperbolic maps.*

DEFINITION. *A rational map is* hyperbolic *if its Julia set and its post-critical set are disjoint.*

From the classification discussion R is hyperbolic if and only if all its critical points are attracted to attracting cycles. The stable dynamics of hyperbolic maps are therefore particularly simple and can be described in terms of the combinatorics of the set of attracting cycles.

We may therefore ask when two maps have similar combinatorial descriptions.

DEFINITION. *Two rational maps R_1 and R_2 are topologically conjugate if there is a homeomorphism $h \colon \hat{\mathbb{C}} \to \hat{\mathbb{C}}$ such that $\phi \circ R_1 \circ \phi^{-1} = R_2$.*

Since the conjugation preserves the attracting set, topologically conjugate maps have similar combinatorics.

One of the main open problems in the field is

CONJECTURE (HYPERBOLIC DENSITY). *Hyperbolic maps are open and dense among all rational maps.*

It is easy to see that the hyperbolic maps are open, but the density has so far proved to be intractible. A partial explanation may be due to a theorem of Rees [15] which states

THEOREM 7.1 (REES). *The set of non-hyperbolic maps has positive measure in the set of all rational maps.*

A corollary of the hyperbolic density conjecture would be that non-hyperbolic maps are rigid; that is, completely determined by the combinatorics of their post-critical sets. It would then follows that the combinatorics of the post-critical set (and hence the stable domains) would determine every rational map up to topological conjugacy.

For quadratic polynomials, the Mandelbrot set parametrizes those whose post-critical set is bounded. Douady and Hubbard [5] proved

THEOREM 7.2 (DOUADY-HUBBARD). *Local connectivity of the Mandelbrot set would imply that the hyperbolic maps are open and dense in the set of all quadratic polynomials.*

A related concept is

DEFINITION. *A rational map R_0 is structurally stable if it is topologically conjugate to all maps R in a neighborhood of R_0.*

We know

THEOREM 7.3 (MANÉ, SAD, SULLIVAN). [11] *In the space of rational maps, the structurally stable maps are open and dense.*

A statement equivalent to the hyperbolic density conjecture is therefore

CONJECTURE (STRUCTURAL STABILITY). *A structurally stable rational map is hyperbolic.*

Another equivalent characterization of hyperbolic maps is

PROPOSITION 7.4. *A rational map R is hyperbolic if and only if it is expanding on its Julia set; that is, there exists $\lambda > 1$ and a neighborhood N of J_R such that*

$$|R'(z)| > \lambda, \quad \forall z \in N.$$

This characterization can be used to prove that the area of the Julia set of a hyperbolic map is either zero or the Julia set is the whole sphere.

CONJECTURE (AREA ZERO). *The Julia set of any rational map either has area zero or is the whole Riemann sphere.*[1]

This is clearly a delicate problem since for maps whose Julia set is not the whole sphere, Sullivan [6], proved that the Hausdorff dimension of the Julia set of a hyperbolic map is strictly less than two but Shishikura [16], recently showed that for a generic c on the boundary of the Mandelbrot set the Hausdorff dimension of the Julia set of $z^2 + c$ is two. McMullen [12] showed that there are entire maps in the family $a \sin z + b$ whose Julia set has positive area.

An excellent discussion of these ideas and the current state of the subject may be found in [13]. The articles of Devaney, Branner,Douady and Hubbard also treat these questions.

8. Quasiconformal mappings

In this section we give a definition and a brief discussion of how the theory of quasiconformal maps fits into the study of rational maps. A good standard reference for the basic theory of quasiconformal mappings is [1].

Holomorphicity imposes a very strong structure on our dynamical systems; although they are maps of the 2 dimensional sphere, they have more of a 1 dimensional flavor then then one might expect because ordinary derivatives are defined.

Quasiconformal maps have some of the same properties as conformal maps but are more flexible. We shall define these maps here in an intuitive but not completely rigorous manner. Recall that at a point z, a holomorphic map f has the property that all its directional derivatives $D_\theta f$ have the same modulus.

DEFINITION. *Let $\phi : \hat{\mathbf{C}} \to \hat{\mathbf{C}}$ be an almost everywhere differentiable homeomorphism. If there exists $K > 1$, such that for every z at which it is differentiable,*

$$\frac{\max_\theta D_\theta \phi}{\min_\theta D_\theta \phi} < K,$$

ϕ *is called* K-*quasiconformal.*

---

[1]added in press: S. van Strien has announced a counter-example. He claims to have a polynomial of the form $z^q + \epsilon$ for a large q and a small $|\epsilon|$ whose Julia set has positive area.

We can associate the measurable function

$$\mu_\phi(z) = \phi_{\bar{z}}/\phi_z$$

to a K-quasiconformal ϕ and $|\mu_\phi(z)| < k = (K-1)/(K+1) < 1$. The *Measurable Riemann Mapping Theorem* proved by Mori, Ahlfors and Bers says the converse is true.

THEOREM 8.1 (MEASURABLE RIEMANN MAPPING THEOREM). *If $\mu(z)$ is a measurable function defined on $\hat{\mathbb{C}}$ such that $\|\mu(z)\|_\infty < k < 1$, there exists a unique quasiconformal homeomorphism ϕ such that $\mu(z) = \mu_\phi(z)$ fixing $\{0, 1, \infty\}$. Moreover, if μ depends holomorphically on parameters, so does ϕ.*

Suppose R is a rational map and ϕ is a quasiconformal homeomorphism of $\hat{\mathbb{C}}$. Then

$$\tilde{R} = \phi \circ R \circ \phi^{-1}$$

is rational if and only if

$$\mu_\phi(R(z))\bar{R}'(z)/R'(z) = \mu_{\phi(z)}.$$

When this condition holds we say that μ_ϕ defines an *invariant line field* for the rational map R and that $\tilde{R}$ is *quasiconformally conjugate* to R.

In Sullivan's proof of the no-wandering domain theorem, he showed that if a wandering domain existed for a rational map R, it would support invariant line fields. In fact, there would be so many line fields that the space of rational maps quasiconformally conjugate to R would be an infinite dimensional vector space. Since R depends only on its finitely many coefficients this is impossible.

One of the reasons hyperbolic maps are easy to work with is that it is easy to construct invariant line fields supported on their stable sets. Since their Julia sets have zero measure, we obtain all quasiconformally conjugate maps from these invariant line fields. Using these ideas one can prove ([**17**]):

THEOREM 8.2. *If two hyperbolic maps are topologically conjugate they are quasiconformally conjugate.*

An open problem related to the hyperbolic density conjecture is then

CONJECTURE. *If two maps are topologically conjugate they are quasiconformally conjugate.*

Since quasiconformal maps have more structure than topological maps we would obtain a little more control of structurally stable maps if this conjecture were proved.

The area zero conjecture for rational maps has an analog in the theory of Kleinian groups: the limit set of a finitely generated Kleinian group either has area zero or is the whole sphere.

In [6], Sullivan came close to proving this conjecture for Kleinian groups by showing that the limit set of a finitely generated Kleinian group supports no invariant line field. The analogous conjecture for rational maps is:

CONJECTURE (NO INVARIANT LINE FIELDS). *The Julia set of a rational map supports no invariant line field.*

This conjecture implies the hyperbolic density conjecture (see [13]). Recently, progress has been made on this conjecture for quadratic polynomials, and in particular for quadratic polynomials with real c, see [13, 18, 19].

The author would like to thank Scott Sutherland for providing the computer pictures for this article.

REFERENCES

1. L. Ahlfors. *Lectures on Quasiconformal Mappings.* Van Nostrand, 1966.
2. A. F. Beardon. *Iteration of Rational Functions.* Springer, New York, 1991.
3. P. Blanchard. Complex analytic dynamics on the Riemann sphere. *Bull. A. M. S.*, 11:85–141, 1984.
4. L. E. Böttcher. The principal laws of convergence of iterates and their applications to analysis. *Izv. Kazan. Fix.-Mat. Obshch*, 14:155–234, 1904. Russian.
5. A. Douady and J. H. Hubbard. *Étude dynamique des polynômes complexes, Parts I,II.* Publ. Math. Orsay, 1984-85.
6. D.P.Sullivan. Conformal dynamical systems. In *Geometric Dynamics*, volume 1007, pages 725–752. Springer-Verlag Lecture notes, 1983.
7. P. Grzegorczyk, F. Przytycki, and W. Szlenk. On iterations of Misiurewicz's rational maps on the Riemann sphere. Inst. of Math. Polish Academy of Sciences Šniadeckich 8, preprint.
8. T. Gamelin and L. Carleson. *Complex Dynamics.* Springer, New York, 1993.
9. L. Keen. Julia sets. In L. Keen and R. Devaney, editors, *Chaos and Fractals: The Mathematics Behind the Computer Graphics*, chapter Julia Sets, pages 57–74. AMS, 1989.
10. G. Koenigs. Recherches sur les intégrales de certaines équations fonctionnelles. *Ann. Sci. Éc. Norm. Sup., Paris*, 1:Supplément 3–41, 1884.
11. R. Mané, P. Sad, and D. P. Sullivan. On the dynamics of rational maps. *Ann. Sci. Éc. Norm. Sup., Paris*, 16:193–217, 1983.
12. C. T. McMullen. Area and Hausdorff dimension of Julia sets of entire functions. *Trans. Amer. Math. Soc.*, 300:329–342, 1987.
13. C. T. McMullen. Frontiers in complex dynamics. *Bull. A. M. S.*, 1994. To appear.
14. J. Milnor. Dynamics in one complex variable: Introductory lectures. MSI-SUNY Stonybrook preprint series.
15. M. Rees. Positive measure sets of ergodic rational maps. *Ann. Sci. Éc. Norm. Sup., Paris*, 19:383–407, 1986.
16. M. Shishikura. The hausdorff dimension of the boundary of the Mandelbrot set and Julia sets. preprint.
17. D. P. Sullivan and C. McMullen. Quasiconformal homeomorphisms and dynamics III: The Teichmüller space of a holomorphic dynamical system. in preparation.
18. G. Świątek. Hyperbolicity is dense in the real quadratic family. Stonybrood IMS preprint 1992/10.
19. J.C. Yoccoz. Sur la connexité locale des ensembles de Julia et du connexité des polynômes. in preparation.
20. J.C. Yoccoz. Theoreme de Siegel, polynomes quadratiques et nombres de Brjuno. preprint.

MATHEMATICS DEPARTMENT, CUNY LEHMAN COLLEGE, BRONX, NY 10468, U.S.A.
E-mail address: ljklc@@cunyvm.cuny.edu

Proceedings of Symposia in Applied Mathematics
Volume **49**, 1994

Does a Julia set depend continuously on the Polynomial?

Adrien Douady

Introduction

Given a complex polynomial $f \colon \mathbf{C} \to \mathbf{C}$ of degree $d \geq 2$, one defines the *filled Julia set* $K(f)$ and the actual *Julia set* $J(f)$ (§4). Both are non empty compact sets in $\mathbf{C}$. The set $J(f)$ is the boundary of $K(f)$, and $K(f)$ is the union of $J(f)$ with the bounded connected components of $\mathbf{C} - J(f)$. The set $K(f)$ may have empty interior — in which case $J(f) = K(f)$ — or non empty interior. The set $J(f)$ is the closure of the set of repelling points. Components of the interior of $K(f)$ are always related to non repelling periodic points for f. Periodic points are described in §3.

In this paper, we examine the extent to which extent $K(f)$ and $J(f)$ depend continuously on f, when f ranges in the set $\mathcal{P}_d$ of polynomials of degree d.

We first have to give a meaning to the question. This is done by defining the *Hausdorff metric* on the set $\mathrm{Comp}^*(\mathbf{C})$ of non-empty compact sets in $\mathbf{C}$ (§1). The Hausdorff distance is defined as a sup of two semi-distances, thus continuity for maps in $\mathrm{Comp}^*(\mathbf{C})$ can be decomposed into upper and lower semi-continuity (§2). In contrast with what occurs for maps in $\mathbf{R}$, these two play very different roles: upper semi-continuity is much more natural.

The map $f \mapsto K(f)$ is upper semi-continuous and $f \mapsto J(f)$ is lower semi-continuous. Both are continuous at f_0 if $K(f_0)$ has empty interior (§5).

Having non empty interior for $K(f_0)$ may be caused by the presence of attracting cycles, of parabolic cycles or of Siegel discs. Several of these may occur for the same f_0 if $d > 2$, but in degree 2 they are exclusive of each other. We examine the effect on continuity of these three types of cycles.

1980 *Mathematics Subject Classification (1985 Revision)*. Primary 58F13; Secondary 58F14.

Attracting cycles actually cause no discontinuity (§6). Siegel discs cause a discontinuity for $f \mapsto J(f)$, but none for $f \mapsto K(f)$ (§7).

A parabolic cycle causes discontinuity in both $f \mapsto K(f)$ and $f \mapsto J(f)$. These discontinuities can be precisely described and analyzed. We shall do that in Part II for the simplest example, the map $z \mapsto z + z^2$. The main tool for this study is provided by the *Fatou coordinates*.

I want to thank Bodil Branner and Núria Fagella for their help during the preparation of this text, Jacques Carette for providing the computer pictures, and Dan Sorensen for the colour plates. I want to thank also Robert Devaney, Boston University, and the American Mathematical Society for the organization of this meeting, with a special thought of sympathy for Carole Kohansky.

This text is being issued in French in the proceedings of a similar meeting (rencontres X-UPS) organized by Nicole Berlineand C. Sabbagh, the Ecole Polytechnique, and the Union des Prefesseurs de Spéciale.

I Continuity Properties

1 The Hausdorff metric.

Let E be a metric space. We denote by $\mathrm{Comp}(E)$ the set of compact subsets of E and we set $\mathrm{Comp}^*(E) = \mathrm{Comp}(E) - \{\emptyset\}$. We shall provide $\mathrm{Comp}^*(E)$ with a distance called the Hausdorff distance. We give the definition and some properties in the general setting of arbitrary metric spaces, but actually we are only interested in the case $E = \mathbf{C}$.

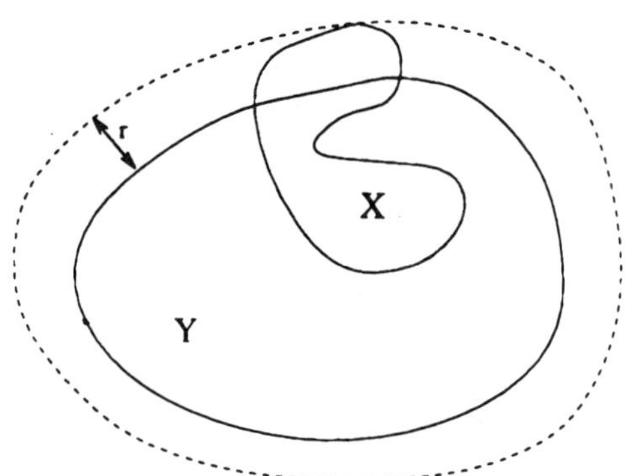

Let X and Y be two compact subsets of E. We say that X *is contained in Y up to r* if X is contained in the r-neighborhood of Y, i.e. if $d(x, Y) \leq r$ for any $x \in X$. We denote by $\partial(X, Y)$ the smallest r such that $X \subset Y$ up to r, i.e.

$$\partial(X, Y) = \sup_{x \in X} \ d(x, Y).$$

The *Hausdorff distance d_H* is defined by

$$d_H(X, Y) = \sup(\partial(X, Y), \partial(Y, X)).$$

Note that the usual conventions concerning the empty set lead us to set $d(x, \emptyset) = \infty$, $\partial(X, \emptyset) = \infty$ for $X \neq \emptyset$, $\partial(\emptyset, \emptyset) = 0$. In order to avoid irrelevant difficulties, we shall work on $\mathrm{Comp}^*(E)$, i.e. we consider only non-empty compact subsets of E.

Proposition 1.1 - *The function d_H is a distance on $\mathrm{Comp}^*(E)$.*

Lemma 1.2 - *(a) We have $\partial(X, Y) = 0$ iff $X \subset Y$.*
(b) For X, Y, Z in $\mathrm{Comp}^*(E)$, we have

$$\partial(X, Z) \leq \partial(X, Y) + \partial(Y, Z).$$

Proof: (a) We have
$\partial(X, Y) = 0 \Leftrightarrow (\forall x \in X) \ d(x, Y) = 0 \Leftrightarrow (\forall x \in X) \ x \in Y \Leftrightarrow X \subset Y$.
We have the second equivalence because Y is closed.

(b) Set $r_1 = \partial(X, Y)$ and $r_2 = \partial(Y, Z)$. For $x \in X$, there is a $y \in Y$ such that $d(x, y) \leq r_1$, and then one can find $z \in Z$ such that $d(y, z) \leq r_2$. So $d(x, z) \leq r_1 + r_2$ and $d(x, Z) \leq r_1 + r_2$. Since this holds for any $x \in X$, we have $\partial(X, Z) \leq r_1 + r_2$.

q.e.d.

Proof of Prop. 1.1: (a) $d_H(X, Y) = d_H(Y, X)$ is obvious from the definition.

(b) $d_H(X, Y) = 0 \Leftrightarrow \partial(X, Y) = \partial(Y, X) = 0 \Leftrightarrow (X \subset Y$ and $Y \subset X) \Leftrightarrow X = Y$.

(c) from $\partial(X, Z) \leq \partial(X, Y) + \partial(Y, Z)$
and $\partial(Z, X) \leq \partial(Y, X) + \partial(Z, Y)$,
we get $d_H(X, Z) \leq d_H(X, Y) + d_H(Y, Z)$.

q.e.d.

From now on, we provide $\mathrm{Comp}^*(E)$ with the Hausdorff distance and thus consider it as a metric space. This allows us to make use of expressions such as "$X_n \to X$ when $n \to \infty$", or "X_λ depends continuously on $\lambda \in \Lambda$" (when Λ is a topological space).

2. Semi continuous mappings into $\mathrm{Comp}^*(E)$.

The Hausdorff distance $(X, Y) \mapsto d_H(X, Y)$ is the sup of two semi-distances $(X, Y) \mapsto \partial(X, Y)$ and $(X, Y) \mapsto \partial(Y, X)$. This enables us to decompose the property of continuity for a map $\Lambda \to \mathrm{Comp}^*(E)$ into upper and lower semi-continuities.

Compare with $\mathbf{R}$: the distance $(x, y) \mapsto |y - x|$ on $\mathbf{R}$ can be expressed as

$$|y - x| = \sup((y - x)^+, \ (x - y)^+)$$

where $r^+ = \sup(r, 0)$. Recall that a map $f \colon \Lambda \to \mathbf{R}$ is said to be *upper semi-continuous* if

$$(f(\lambda) - f(\lambda_0))^+ \to 0 \text{ when } \lambda \to \lambda_0,$$

i.e.

$$(\forall \epsilon > 0) \ (\exists V, \ nbd. \ of \ \lambda_0) \ (\forall \lambda \epsilon V), \ \ f(\lambda) < f(\lambda_0) + \epsilon,$$

and *lower semi-continuous* if

$$(f(\lambda_0) - f(\lambda))^+ \to 0 \text{ when } \lambda \to \lambda_0,$$

i.e.

$$(\forall \epsilon > 0) \ (\exists V, \ nbd. \ of \ \lambda_0) \ (\forall \lambda \in V), \ f(\lambda) > f(\lambda_0) - \epsilon.$$

By analogy, we say that a map $\phi \colon \Lambda \to \mathrm{Comp}^*(E)$ is *upper semi-continuous* if

$$\partial(\phi(\lambda), \ \phi(\lambda_0)) \to 0 \text{ when } \lambda \to \lambda_0$$

i.e.

$$(\forall \epsilon > 0) \ (\exists V, \ nbd. \ of \ \lambda_0) \ (\forall \lambda \in V), \ \ \phi(\lambda) \subset \phi(\lambda_0) \text{ up to } \epsilon,$$

and *lower semi-continuous* if

$$\partial(\phi(\lambda_0), \ \phi(\lambda)) \to 0 \text{ when } \lambda \to \lambda_0,$$

i.e.

$$(\forall \epsilon > 0) \ (\exists V, \ nbd. \ of \ \lambda_0) \ (\forall \lambda \in V), \ \ \phi(\lambda) \supset \phi(\lambda_0) \text{ up to } \epsilon.$$

Actually, there is a difference with the case of $\mathbf{R}$, which is all favorable to this new case. In the case of $\mathbf{R}$, it is very hard for a normal person, even with some training, to remember which is which and not to get mixed up between the two semi-continuities. In the case of $\mathrm{Comp}^*(E)$, the two semi-continuities play very different roles: upper semi-continuity is a much more natural thing to consider, because of the interpretation given by Prop. 2.1 below, while the lower semi-continuity is kind of a bizarre property.

For E and F two metrizable topological spaces, a map $f: E \to F$ is said to be *proper* if it is continuous and the inverse image of a compact set in F is a compact set in E. Any proper map is closed, i.e. the image of a closed set is closed.

Proposition 2.1 *Let Λ and E be two metrizable topological spaces and $(X_\lambda)_{\lambda \in \Lambda}$ be a family of subsets of E. Consider the space $\mathcal{X}$ of pairs $(\lambda, x) \in \Lambda \times E$ such that $x \in X_\lambda$. Then the following properties are equivalent:*

(1) The map $\lambda \mapsto X_\lambda$ is an upper semi-continuous map from Λ into $\mathrm{Comp}^*(E)$;

(2) $\mathcal{X}$ is closed in $\Lambda \times E$ and the projection $p_{\mathcal{X}}: \mathcal{X} \to \Lambda$ is proper and surjective.

Proof: $(2) \Rightarrow (1)$: For each λ, the set X_λ is a non empty compact set in E. Given $\lambda_0 \in \Lambda$ and $\epsilon > 0$, the set $A = \{(\lambda, x) \in \mathcal{X} \, | \, d_H(x, X_{\lambda_0}) \geq \epsilon\}$ is a closed subset of $\mathcal{X}$ not intersecting $p_{\mathcal{X}}^{-1}(\lambda_0)$. So $p_{\mathcal{X}}(A)$ is closed in Λ and $W = \Lambda - \rho_{\mathcal{X}}(A)$ is a neighborhood of λ_0. For $\lambda \in W$, we have $\partial(X_\lambda, X_{\lambda_0}) < \epsilon$.

$(1) \Rightarrow (2)$: Since $X_\lambda \neq \emptyset$ for each λ, the map $p_{\mathcal{X}}$ is surjective. Let $L \subset \Lambda$ be a compact set, and let $((\lambda_n, x_n))_{n \in \mathbf{N}}$ be a sequence in $p_{\mathcal{X}}^{-1}(L)$. Since (λ_n) is a sequence in L, we can suppose it is converging to a point $\lambda_0 \in L$. Set $r_n = d(x_n, X_{\lambda_0})$. By upper semi-continuity, $r_n \leq \partial(X_{\lambda_n}, X_{\lambda_0}) \to 0$. For each n, one can find a point $y_n \in X_{\lambda_0}$ such that $d(x_n, y_n) = r_n$. From $(y_n)_{n \in \mathbf{N}}$ we can extract a sequence converging to a point $y_0 \in X_{\lambda_0}$. Then $(\lambda_n, x_n) \to (\lambda_0, y_0)$.

q.e.d.

Remark 2.2: If Λ is locally compact, condition (2) is equivalent to: $(2')$ $\mathcal{X}$ is closed in $\Lambda \times E$ and, for each $\lambda_0 \in \Lambda$ there is a neighborhood V of λ_0 in Λ and a compact set K in E such that $X_\lambda \subset K$ for all $\lambda \in V$.

3. Periodic points

3.1 Multiplier.

Let f be a complex polynomial of degree d in one variable, viewed as a map $f: \mathbf{C} \to \mathbf{C}$. We denote by f^k the k-th iterate of f.

A point $x \in \mathbf{C}$ is *periodic of period k* if $f^k(x) = x$, $f^i(x) \neq x$ for $0 < i < k$. The *cycle* spanned by x is then $\xi = \{x_0, \ldots, x_{k-1}\}$, where $x_i = f^i(x)$. The *multiplier* of ξ is

$$\rho_\xi = f'(x_0) \cdot f'(x_1) \cdot \ldots \cdot f'(x_{k-1}) = (f^k)'(x_i)$$

for all i. If z is close to x_i, then $f^k(z) - x_i$ is approximately $\rho_\xi \cdot (z - x_i)$.

The cycle ξ is called *attracting* if $|\rho_\xi| < 1$, *repelling* if $|\rho_\xi| > 1$, *indifferent* (or *neutral*) if $|\rho_\xi| = 1$. Indifferent cycles are divided into *parabolic cycles* (case where ρ_ξ is a root of unity) and *irrational indifferent* cycles.

The multiplier of a periodic point is the multiplier of the cycle it generates. A periodic point is attracting, repelling, etc., if the cycle it generates is such.

3.2 Attracting cycles.

If x is an attracting periodic point of period k, the *basin A_x* of x is $\{z \in \mathbf{C} \,|\, f^{nk}(z) \to x$ when $n \to \infty\}$. It is an open set contained in $\overset{\circ}{K}(f)$, and $x \in A_x$. The *immediate basin A_x^o* is the connected component of A_x which contains x. The *basin* (resp. the *immediate basin*) of the cycle ξ spanned by x is

$$A_\xi = \bigcup_{0 \leq i < k} A_{x_i} \quad (\text{resp. } A_\xi^o = \bigcup_{0 \leq i < k} A_{x_i}^o).$$

The following result is classical. Recall that a *critical point* for f is a point where the derivative f' vanishes.

Theorem 3.1 (Fatou, Julia) – *The immediate basin of an attracting cycle always contains a critical point.*

Corollary. - *The maximum number of attracting cycles for a polynomial of degree d is $d - 1$.*

Indeed, a polynomial of degree d has at most $d - 1$ distinct critical points.

In fact, there is a better result:

Theorem 3.2 - *The maximum number of non repelling cycles for a polynomial of degree d is $d - 1$.*

Sketch of proof: The proof relies on the notion of polynomial-like mappings. A *polynomial-like mapping $f : U' \to U$*, where U and

U' are two open sets in $\mathbf{C}$ bounded by Jordan curves γ and γ' with $\overline{U'} \subset U$, is a holomorphic map $U' \to U$ which can be extended into a continuous map $\overline{U'} \to \overline{U}$, inducing a map $\gamma' \to \gamma$. The *degree* of f is the degree of the induced map $\gamma' \to \gamma$. A polynomial-like mapping of degree d has $d-1$ critical points (counting with multiplicity). If f is a polynomial of degree d, the map $f \colon f^{-1}(D_R) \to D_R$ is polynomial-like of degree d for R great enough.

Theorem 3.1 can be extended to polynomial-like mappings: each attracting cycle attracts at least one critical point. The proof can be copied on the proof for polynomials. As a consequence, a polynomial-like mapping of degree d has at most $d - 1$ attracting cycles.

The advantage of this new setting is the following: while it is not easy to modify a polynomial so as to make all non repelling cycles attracting (without changing the degree), this is very easy for polynomial-like mappings. Therefore a polynomial-like mapping of degree d has at most $d - 1$ non repelling cycles, else the modified map would give a contradiction. And this holds in particular for polynomials.

q.e.d.

3.3 Parabolic cycles

Suppose now that x is a parabolic point of period k and multiplier $\rho = e^{2\pi i p/q}$ (p, q relatively prime). Then f^{kq} is tangent to the identity at x. One can find Q half lines of origin x called the *attracting axes*, alternating with Q others called the *repelling axes*, where Q is a multiple of q with the following property: if z is close to x on an attracting axis (resp. on a repelling axis) then f^{kq} moves z towards x (resp. away from x).

The attracting and repelling axes are determined in the following way. Placing the origin at x, the map f^{kq} takes the form

$$z \mapsto z(1 + cz^Q + O(z^{Q+1}))$$

for some $c \neq 0$, $Q \geq 1$. The values of z for which cz^Q is real < 0 (resp. > 0) form the attracting axes (resp. the repelling axes).

The linear tangent map

$$z \mapsto \rho \cdot z$$

maps repelling axes to repelling axes and attracting axes to attracting axes. This is how we know that Q has to be a multiple of q.

For each attracting axis L, there are points z such that $f^{kqn}(z) \to x$ tangentially to L. These points form a non-empty open set A_L

which contains a segment S of L with x as one of its extremities. We call A_L the *basin* of L and the connected component A_L° of A_L which contains S the *immediate basin*. The point x belongs to the boundary of A_L°. The basin A_x (resp. the immediate basin A_x°) of x is the union of A_L (resp A_L°) for L an attracting axis at x, and if ξ is the cycle spanned by x we define the basin (resp. the immediate basin) of ξ as the union of the basins (resp. immediate basins) of the points of the cycle.

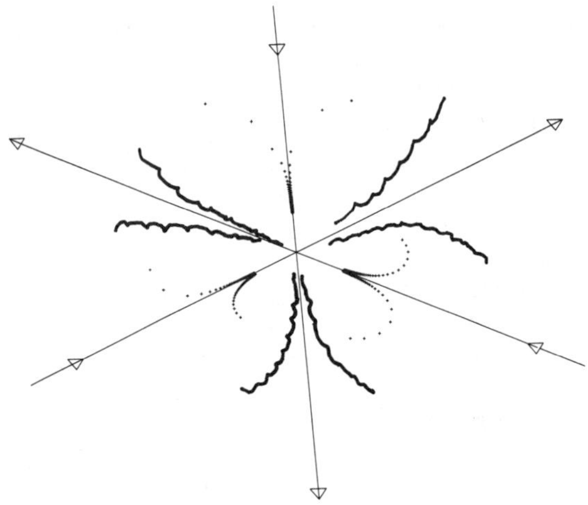

Similar to Theorem 3.1, we have:

Theorem 3.3 (Fatou, Julia). *The immediate basin of a parabolic cycle always contains a critical point* (in fact at least $\nu = Q/q$ critical points).

3.4 Irrational indifferent cycles.

Suppose now that x is an irrational indifferent periodic point, with period k and multiplier $\rho = e^{2\pi i\theta}$, $\theta \in \mathbf{R} - \mathbf{Q}$. Then x may be linearizable (Siegel case) or non-linearizable (Cremer case). We say that x is *linearizable* if there is an open neighborhood U of x such that $f^k(U) = U$ and an isomorphism $\phi: U \to D$ such that $\phi \circ f^k \circ \phi^{-1}: D \to D$ is the rotation $z \mapsto \rho \cdot z$. The biggest possible U is called the *Siegel disc Δ_x* of x.

Results by Cremer, Siegel, Bruno, Yoccoz give conditions for an irrational indifferent point to be linearizable. For quadratic polynomials, we have

Theorem 3.4 (Bruno, Yoccoz) - *Let f be a quadratic polynomial having an irrational indifferent periodic point with period k and multiplier $\rho = e^{2\pi i\theta}$, $\theta \in \mathbf{R} - \mathbf{Q}$. Let $\frac{p_n}{q_n}$ be the approximants of θ given by the continued fraction expansion. Then x is linearizable if and only if*

$$(\mathcal{B}) \quad \sum \frac{\log q_{n+1}}{q_n} < \infty.$$

The implication $(\mathcal{B}) \Rightarrow$ linearizability, is due to Brjuno, and is valid for a polynomial of any degree – in fact for any germ of holomorphic mapping. The converse is due to Yoccoz.

4. Filled Julia set and Julia set of a Polynomial

Let $f \colon \mathbf{C} \to \mathbf{C}$ be a complex polynomial of degree $d \geq 2$. We denote f^n the n-th iterate of f, i.e. $f^0 = \mathrm{Id}$, $f^{n+1} = f \circ f^n$. We sometimes write z_n for $f^n(z)$.

We say that z *escapes* under f if $f^n(z) \to \infty$ when $n \to \infty$. There is a $R > 0$ such that any z with $|z| \geq R$ escapes. Such a value of R is called an *escape radius*. For $f \colon z \mapsto a_d z^d + \ldots + a_0$, the radius

$$R = \frac{1 + |a_d| + \ldots + |a_0|}{|a_d|}$$

is an escape radius.

Actually, as soon as a point starts escaping for good, the convergence to ∞ is extremely rapid. As an example, for $f(z) = z^2 + c$ with $|c| \leq 2$, if $|z_n| \geq 3$ for some n, then $|z_{n+10}| \geq 10^{300}$: it is bigger than the ratio of the volume of the known universe, way to the most remote quasars observed, to the volume of a proton.

We denote by $K(f)$ the set of points which do not escape under f, and we call it the *filled Julia set* of f. The actual *Julia set* is the boundary $J(f)$ of $K(f)$. The set $K(f)$ is compact. Indeed, it is contained in D_R if R is an escape radius, and it is closed: if a point z escapes, then $|f^n(z)| > R$ for some n, thus $|f^n(z')| > R$ with the same n if z' is sufficiently close to z, so the set of escaping points is open.

The set $K(f)$ is never empty, since it contains the fixed points, solutions of $f(z) - z = 0$. In fact it is always infinite non-countable. It follows from the maximum principle that its complement $\mathbf{C} - K(f)$

cannot have a bounded component, thus it is simply connected. According to the coefficients of f the set $K(f)$ may be connected or not, and its interior $\overset{\circ}{K}(f)$ may be empty (in which case $J(f) = K(f)$) or not empty. By a theorem of Fatou and Julia, $K(f)$ is connected if and only if none of the critical points escapes.

Proposition 4.1 (a) *Any attracting periodic point lies in $\overset{\circ}{K}(f)$.*

   (b) Any repelling periodic point lies in $J(f)$.

   (c) Any parabolic point lies in $J(f)$.

   (d) Any Siegel point (i.e. linearizable irrational indifferent periodic point) lies in $\overset{\circ}{K}(f)$.

   (e) Any Cremer point (i.e. non linearizable irrational indifferent periodic point) lies in $J(f)$.

Proof: (a) and (d) are immediate: the attracting basin of x or the Siegel disc is a neighborhood of x contained in $K(f)$.

Conversely, suppose x is periodic of period k and $x \in \overset{\circ}{K}(f)$, and let U be the connected component of $\overset{\circ}{K}(f)$ containing x. Then f^k maps U to U. The open set U is isomorphic to $\mathbf{D}$, so let $\phi \colon U \to \mathbf{D}$ be an isomorphism with $\phi(x) = 0$ and consider $h = \phi \circ f^k \circ \phi^{-1} \colon \mathbf{D} \to \mathbf{D}$. We have $h(0) = 0$ and $h'(0)$ is the multiplier ρ. By Schwarz lemma, $|\rho| \leq 1$, and if $|\rho| = 1$ the map h is the rotation $z \mapsto \rho \cdot z$. This proves (b) and (e).

Note that, if x is parabolic with $\rho = e^{2\pi i p/q}$, then f^{kq} is tangent to the identity at x, but cannot be the identity since it is a polynomial of degree dkq. So x cannot be linearizable and (c) follows.

$$\text{q.e.d.}$$

By a theorem of Sullivan and former results of Fatou, components of the interior of $K(f)$ are always related to nonrepelling cycles in a way that we shall describe.

   The sets of the following form

   (A) Immediate basins of attracting periodic points;

   (P) Immediate basins of attracting axes of parabolic points;

   (S) Siegel disks

are connected components of $\overset{\circ}{K}(f)$.

   If U is a connected component of $\overset{\circ}{K}(f)$, its image $f(U)$ is again a connected component of $\overset{\circ}{K}(f)$. We say that U is a periodic component

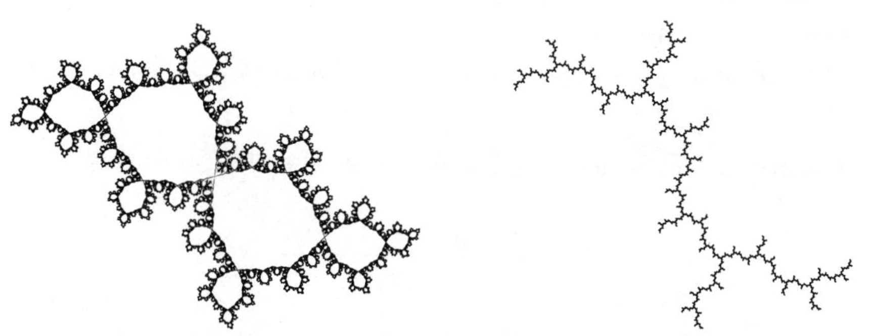

Two with Attracting Cycles

Two with parabolic cycles

One with a Siegel disk $\overset{\circ}{K} = \emptyset$: all cycles are repelling

Some examples of quadratic Julia sets.

if $f^k(U) = U$ for some $k > 0$, preperiodic if $f^l(U)$ is periodic for some $l \geq 0$. Components of type (A),(P),(S) above are periodic.

Theorem 4.1 (Sullivan) - *Each connected component of $\overset{\circ}{K}(f)$ is preperiodic.*

Theorem 4.2 (Fatou) - *Each periodic component of $\overset{\circ}{K}(f)$ is of one of the types*
(A),(P),(S).

Corollary 4.3 - *If all cycles of f are repelling, then $\overset{\circ}{K}(f) = \emptyset$.*

The set $J(f)$, boundary of $K(f)$, is compact, non empty, and has empty interior. The set $K(f)$ is the union of $J(f)$ with the bounded components of $\mathbf{C} - J(f)$. By a theorem of Fatou and Julia, $J(f)$ is the closure of the set of repelling periodic points of f, and for any $\alpha \in J(f)$, the set $J(f)$ is the closure of the set of iterated preimages of α.

5. Semi-continuity of $K(f)$ and $J(f)$ as functions of f

Let $\mathcal{P}_d$ denote the set of complex polynomials of degree d. Such a polynomial can be written

$$f(z) = a_d z^d + a_{d-1} z^{d-1} + \ldots + a_0, \text{ with } a_d \neq 0.$$

So $\mathcal{P}_d$ can be identified with $\mathbf{C}^* \times \mathbf{C}^d$.

Theorem 5.1 - *(a) The map $f \mapsto K(f)$ from $\mathcal{P}_d$ to $\mathrm{Comp}^*(\mathbf{C})$ is upper semi-continuous.*
(b) the map $f \mapsto J(f)$ from $\mathcal{P}_d$ to $\mathrm{Comp}^*(\mathbf{C})$ is lower semicontinuous.

Proof: (a) For each $f \in \mathcal{P}_d$, consider the escape radius

$$R_f = \frac{1 + |a_d| + \ldots + |a_0|}{|a_d|}.$$

It depends continuously on f. The set of pairs (f, z) in $\mathcal{P}_d \times \mathbf{C}$ such that z escapes under f is $\{(f, z) | (\exists n) \ |f^n(z)| > R_f\}$. So it is open and $\mathcal{K} = \{(f, z) | z \in K(f)\}$ is closed in $\mathcal{P}_d \times \mathbf{C}$. Since $K(f) \neq \emptyset$ for all f and $\mathcal{K} \subset \{(f, z) \ | |z| \leq R_f\}$, the projection $\pi_\mathcal{K}: \mathcal{K} \to \mathcal{P}_d$ is proper and surjective (cf. Remark (2.2)). So we can apply Prop. 2.1.

(b) We shall use the description of $J(f)$ as the closure of the set of repelling periodic points. Take $f_0 \in \mathcal{P}_d$ and $\epsilon > 0$. One can find a finite set $X = \{x_1, \ldots, x_N\}$ of repelling periodic points for f_0 filling $J(f_0)$ up to $\frac{\epsilon}{2}$, i.e. such that $\partial(J(f_0), X) \leq \frac{\epsilon}{2}$. Each x_i is a solution of an equation $f_0^{k_i}(z) - z = 0$. It is a simple solution of this equation because the derivative is $\rho_{x_i} - 1 \neq 0$. By the implicit function theorem one can find a neighborhood W of f_0 in $\mathcal{P}_d$ and holomorphic functions $\xi_i \colon W \to \mathbf{C}$ such that $\xi_i(f)$ is periodic of period k_i for f, with $\xi_i(f_0) = x_i$. If W has been chosen small enough, $\xi_i(f)$ is repelling, thus belongs to $J(f)$, and $|\xi_i(f) - x_i| \leq \frac{\epsilon}{2}$ for all $f \in W$. Then $\partial(X, J(f)) \leq \partial(X, \{x_i(f)\}) \leq \epsilon/2$ and $\partial(J(f_0), J(f)) \leq \epsilon$.

q.e.d.

Corollary 5.2 *Let $f_0 \in \mathcal{P}_d$ be a point such that $\overset{\circ}{K}(f_0) = \emptyset$, so that $J(f_0) = K(f_0)$. Then the two maps $f \mapsto K(f)$ and $f \mapsto J(f)$ are continuous at f_0.*

This follows from the theorem and Lemma 5.4 below. We first prove Lemma 5.3, not in order to use it but just to put ourselves in the mood.

Lemma 5.3 - *Let Λ be a topological space, u and v two functions: $\Lambda \to \mathbf{R}$ with u lower semi-continuous and v upper semi-continuous. Suppose $u(\lambda) \leq v(\lambda)$ for all $\lambda \in \Lambda$ and $u(\lambda_0) = v(\lambda_0)$. Then u and v are continuous at λ_0.*

Proof: Take $\epsilon > 0$. There is a neighborhood W of λ_0 such that, for $\lambda \in W$, we have $u(\lambda) \geq u(\lambda_0) - \epsilon$ and $v(\lambda) \leq v(\lambda_0) + \epsilon$. Then, for $\lambda \in W$, we have $u(\lambda_0) - \epsilon \leq u(\lambda_0) \leq v(\lambda) \leq v(\lambda_0) + \epsilon$, thus $|u(\lambda) - u(\lambda_0)| \leq \epsilon$ and $(v(\lambda) - v(\lambda_0)) \leq \epsilon$.

q.e.d.

Lemma 5.4 - *Let Λ be a topological space, $\lambda \mapsto X_\lambda$ and $\lambda \mapsto Y_\lambda$ two maps $\Lambda \to \mathrm{Comp}^* E$, the first one lower semi-continuous and the second one upper semi-continuous. Suppose $X_\lambda \subset Y_\lambda$ for each $\lambda \in \Lambda$ and $X_{\lambda_0} = Y_{\lambda_0}$. Then $\lambda \mapsto X_\lambda$ and $\lambda \mapsto Y_\lambda$ are continuous at Λ_0.*

Proof: Take $\epsilon > 0$. There is a neighborhood W of λ_0 in Λ such that $\partial(X_{\lambda_0}, X_\lambda) \leq \epsilon$ and $\partial(Y_\lambda, Y_{\lambda_0}) \leq \epsilon$ for $\lambda \in W$. For $\lambda \in W$, we have also

$$\partial(Y_{\lambda_0}, Y_\lambda) = \partial(X_{\lambda_0}, Y_\lambda) \leq \partial(X_{\lambda_0}, X_\lambda) \leq \epsilon$$

and

$$\partial(X_\lambda, X_{\lambda_0}) = \partial(X_\lambda, Y_{\lambda_0}) \leq \partial(Y_\lambda, Y_{\lambda_0}) \leq \epsilon.$$

q.e.d.

6. The effect of attracting basins.

Let us denote by $A(f)$ the union of the basins of the attracting cycles of f. The set $A(f)$ is a union of connected components of $\overset{\circ}{K}(f)$. Therefore it is open and $J(f) \cup A(f)$ is closed (thus compact).

Proposition 6.1 - *(a) The map $f \mapsto J(f) \cup A(f)$ is a lower semi-continuous map $\mathcal{P}_d \to \mathrm{Comp}^*(\mathbf{C})$.*

(b) The map $f \mapsto K(f) - A(f)$ is an upper semi-continuous map $\mathcal{P}_d \to \mathrm{Comp}^*(\mathbf{C})$.

Corollary 6.2 - *If the interior of $K(f_0)$ consists only in attracting basins, then the two maps $f \mapsto K(f)$ and $f \mapsto J(f)$ are continuous at f_0.*

Lemma 6.3 - *The set $\mathcal{A} = \{(f,z) \in \mathcal{P}_d \times \mathbf{C} \,|\, z \in A_f\}$ is open in $\mathcal{P}_d \times \mathbf{C}$.*

Before we prove this lemma, we introduce a notion.

We say that a set $L \subset \mathbf{C}$ is a *collapsing trap* for f of period dividing k if L is a compact convex non-empty set with $f^k(L) \subset \overset{\circ}{L}$ and $|(f^k)'| < 1$ on L. If L is a collapsing trap for f of period dividing k, there is a point $x \in L$ such that $f^{nk}(z) \to x$ for any $z \in L$. The point x is an attracting periodic point for f and L is contained in its immediate basin.

If x is an attracting periodic point for f, the closed disc $\overline{D}_{x,r}$ is a collapsing trap for f if r is sufficiently small. If L is a collapsing trap for f_0, it is also a collapsing trap for f if f is sufficiently close to f_0.

Proof of Lemma 6.3: Take (f_0, z_0) in $\mathcal{A}$. One can find a collapsing trap L for f_0 and an integer n such that $f_0^n(z_0) \in \overset{\circ}{L}$. For (f, z) sufficiently close to (f_0, z_0), we have $f^n(z) \in L$ with the same n, and L is a collapsing trap for f, so $z \in A(f)$.

q.e.d.

Proof of Prop. 6.1: (a) Fix $f_0 \in \mathcal{P}_d$ and $\epsilon > 0$. Let $X = X_1 \cup X_2$ be a finite set in $J(f_0) \cup A(f_0)$, filling it up to $\frac{\epsilon}{2}$, with $X_1 \subset J(f_0)$ and $X_2 \subset A(f_0)$. One can find a neighborhood W of f_0 in $\mathcal{P}_d$ such that, for $f \in W$, $\partial(X_1, J(f)) \leq \frac{\epsilon}{2}$ and $X_2 \subset A(f)$. Then $\partial(X, J(f) \cup A(f)) \leq \frac{\epsilon}{2}$ and $\partial(J(f_0) \cup A(f_0), J(f) \cup A(f)) \leq \epsilon$.

(b): The projection $p_{\mathcal{K}} \colon \mathcal{K} \to \mathcal{P}_d$ is proper and $\mathcal{K} - \mathcal{A}$ is closed in $\mathcal{K}$, so $p_{\mathcal{K}}$ induces a proper map $\mathcal{K} - \mathcal{A} \to \mathcal{P}_d$. This map is surjective because $K(f) - A(f) \supset J(f) \neq 0$ for all f. So we can apply Prop. 2.1.

q.e.d.

Proof of Corollary 6.2: We can apply Lemma 5.4 to the pairs $(f \mapsto J(f) \cup A(f), \; f \mapsto K(f))$ and $(f \mapsto J(f), \; f \mapsto K(f) - A(f))$.

q.e.d.

7. The effect of Siegel disks

It is easy to show that Siegel disks cause a discontinuity in $f \mapsto J(f)$. It is much harder to show that they cause no discontinuity in $f \mapsto K(f)$.

Prop. 7.1 - *Let f_0 be a polynomial of degree d with a linearizable irrational indifferent periodic point x. Then $f \mapsto J(f)$ is not continuous at f_0.*

Proof: Let k be the period of x. The point x is a simple solution of $f_0^k(z) - z = 0$, because the derivative is $\rho_x - 1$. By the implicit function theorem we can find a holomorphic map $\xi: W \to \mathbf{C}$ such that $\xi(f)$ is periodic of period k under f and $\xi(f_0) = x$, with W a neighborhood of f_0. The map $f \mapsto \rho_{\xi(f)} = (f^k)'(\xi(f))$ is holomorphic, and it is not constant – else all polynomials would have an indifferent cycle! So it is an open map, and one can find a sequence $f_n \to f_0$ such that $\xi(f_n)$ is repelling. Then $\xi(f_n) \in J(f_n)$, and $\partial(\{x\}, \; J(f_n)) = d(x, J(f_n)) \to 0$. But $x \notin J(f_0)$, so $\partial(\{x\}, \; J(f_0)) \neq 0$, and $\partial(J(f_n), J(f_0)) \nrightarrow 0$.

q.e.d.

Prop. 7.2 - *Let f_0 be a polynomial of degree d with an irrational indifferent periodic point x having a Siegel disc Δ. Then $\partial(\overline{\Delta}, \; K(f)) \to 0$ when $f \to f_0$ in $\mathcal{P}_d$.*

Let k be the period of x and let $\phi: \Delta \to D$ be an isomorphism conjugating f_0^k to the rotation $z \mapsto e^{2\pi i \theta} z$. For $r < 1$, set $\Gamma_r = \{z \in \Delta \mid |\phi(z)| = r\}$. Given $\alpha > 0$, we say that a sequence $(z_n)_{n \geq 0}$ is an f_0-*orbit up to α* if for all n, $|z_{n-1} \, f(z_n)| \leq \alpha$. We say that (z_n) *covers* a compact set L up to ϵ if $\cup D_{z_n, \epsilon} \supset L$.

Lemma 7.3 - *For any $r < 1$ and any $\epsilon > 0$, there is an $\alpha > 0$ such that:*

(a) *any sequence (z_n) which is an f_0-orbit up to α with $z_0 \in \Gamma_r$ covers Γ_r up to ϵ;*

(b) any sequence (z_n) which is an f_0-orbit up to α, with $z_0 \in \Delta$, $|\phi(z_0)| \le r$ and $z_N \notin \Delta$ or $|\phi(z_N)| > r$ for some N, covers Γ_r up to ϵ.

Proof: (a) Fix $r < 1$ and $\epsilon > 0$. We can suppose $\epsilon < d(\Gamma_r, \mathbf{C} - \Delta)$. One can find an integer m such that, for any $z \in \Gamma_r$ the set $\{f_0^i(z)\}_{0 \le i < m}$ fills Γ_r up to $\frac{\epsilon}{2}$. One can then find an α such that, for any f_0-orbit (z_n) up to α, with $z_0 \in \Gamma_r$, we have

$$|z_i - f_0^i(z_0)| \le \frac{\epsilon}{2} \text{ for } 0 \le i < m.$$

Then $\{z_i\}_{0 \le i < m}$ covers Γ_r up to ϵ. Note that one can find an α which works for all values of r in $[r_1, r_2]$ if $r_1 < r_2 < 1$.

(b) One can find ϵ_1 such that $|r - r'| \le \epsilon_1 \Rightarrow d_H(\Gamma_r, \Gamma_{r'}^k) \le \frac{\epsilon}{2}$, and α_1 such that $|z - z'| \le \alpha_1 \Rightarrow |\phi(z) - \phi(z')| \le \epsilon_1$. By part (a) one can then find an α_2 such that any f_0-orbit up to α_2 with $d(z_0, \Gamma_r) \le \frac{\epsilon}{2}$ covers Γ_r up to ϵ. Take $\alpha \le \inf(\alpha_1, \alpha_2)$. If (z_n) satisfies the hypothesis of (b), $|\phi(z_n)|$ moves by steps smaller than ϵ_1 as long as it is defined. Then there is an n_0 such that $||\phi(z_{n_0})| - r| \le \epsilon_1$ and the orbit of z_{n_0} covers Γ_r up to ϵ.

q.e.d.

Lemma 7.4 - *For $z \in \Delta$, the distance $d(z, K(f))$ tends to 0 when $f \to f_0$.*

Proof: Let y be a repelling periodic point for f_0. One can find a neighborhood W of f_0 in $\mathcal{P}_d$ and analytic maps ξ and $\eta : W \to \mathbf{C}$ such that $\xi(f)$ and $\eta(f)$ are periodic for f, $\xi(f_0) = x$, $\eta(f_0) = y$. If $z = x$, $d(z, K(f)) \le d(z, \xi(f)) \to 0$, so suppose $z \ne x$. We can restrict to the polynomials f such that $z \notin K(f)$. Fix $\epsilon > 0$ and let α correspond to ϵ as in Lemma 7.3. Denote by $[\xi(f), z]_\Delta$ the inverse image by ϕ of the segment $[\phi(\xi(f)), \phi(z)]$. Since $\xi(f) \in K(f)$ and $z \notin K(f)$, there is a point $w \in [\xi(f), z]_\Delta \cap J(f)$, and we can find, arbitrarily close to it, a point $z_0 \in J(f)$ which is an iterated preimage of $\eta(f)$. We can suppose $|\phi(z_0)| < r = |\phi(z)|$. If f is sufficiently close to f_0, the orbit of z_0 under f is an f_0-orbit up to α. We have $|\phi(z_0)| < r$ and for some N we have $z_N = \eta(f) \notin \Delta$ or $|\phi(\eta(f))| > r$. So $\{z_n\}_{n \in \mathbf{N}}$ covers Γ_r up to ϵ. In particular there is an n such that $|z - z_n| < \epsilon$, thus $d(z, K(f)) < \epsilon$.

q.e.d.

Proof of Prop. 7.2: Fix $\epsilon > 0$, and let Z be a finite set in Δ covering $\overline{\Delta}$ up to $\frac{\epsilon}{2}$. One can find a neighborhood W of f_0 in $\mathcal{P}_d$ such that $d(z, K(f)) \leq \frac{\epsilon}{2}$ for $z \in Z$ and $f \in W$. Then $\partial(\overline{\Delta}, K(f)) \leq \epsilon$ for each $f \in W$. q.e.d.

Corollary 7.5 - *Let* $U = \bigcup_n f^{-n}\Delta$. *Then we have* $\partial(J(f_0) \cup U$ *and* $K(f)) \to 0$ *when* $f \to f_0$.

Corollary 7.6 - *Let* $f_0 \in \mathcal{P}_d$ *be a polynomial without parabolic cycle. Then* $f \mapsto K(f)$ *is continuous at* f_0.

Hint: The interior of $\overset{\circ}{K}(f_0)$ consists only of basins of attracting cycles and Siegel discs together with their iterated preimages.

8. A Mañe-Sad-Sullivan Theorem.

When we know that two compact sets are close to each other for the Hausdorff distance, this does not imply that their topology is related: For instane any compact set can be approximated by Cantor sets, or even by finite sets.

The Mañe-Sad-Sullivan theory provides cases where Julia sets–or filled Julia sets–are homeomorphic. Here is one of the main results:

Theorem 8.1. *Let* Λ *be a* **C**-*analytic manifold and* $(f_\lambda)_{\lambda \in \Lambda}$ *a* **C**-*analytic family of polynomials of degree d. Suppose that* Λ *is simply connected and that* (f_λ) *never has a parabolic cycle for any* $\lambda \in \Lambda$. *Choose a point* $\lambda_0 \in \Lambda$. *Then*

(a) one can find, in a unique way, a family of homeomorphisms

$$(\varphi_\lambda = \varphi_{\lambda_0,\lambda} \colon J(f_{\lambda_0}) \to J(f_\lambda))_{\lambda \in \Lambda}$$

such that
  (i) $(\lambda, x) \mapsto \varphi_\lambda(x)$ *is continuous on* $\Lambda \times J(f_{\lambda_0})$.
  (ii) φ_λ *conjugates* f_{λ_0} *to* f_λ, *i.e.* $\varphi_\lambda \circ f_{\lambda_0} = f_\lambda \circ \varphi_\lambda$.

(b) One can extend, for each λ, *the map* φ_λ *to a homeomorphism* $\phi_\lambda \colon K(f_{\lambda_0}) \to K(f_\lambda)$. *There is no uniqueness. One can arrange so that* $\phi_\lambda(x)$ *depends continuously on* (λ, x), *but not always so that* ϕ_λ *conjugates* f_{λ_0} *to* f_λ.

Corollary 8.2. *Let* S *be the closure in* $\mathcal{P}_d$ *of the set of polynomials having an indifferent cycle. If* f_0 *and* f_1 *are in the same connected component of* $\mathcal{P}_d - S$, *then* $J(f_1)$ *is homeomorphic to* $J(f_0)$ *and* $K(f_1)$ *is homeomorphic to* $K(f_0)$.

A proof of Thm. 8.1. can be found in [MSS] or in [D4]. Corollary 8.2 is an immediate consequence: you can find a simply connected open set Λ in $\mathcal{P}_d - S$ containing f_0 and f_1.

Remarks: (1) Fix f_0 in $\mathcal{P}_d - S$. For f close to f_0, the compact set $K(f)$ is the image of $K(f_0)$ by a homeomorphism close to the identity. This is stronger than the conclusion of Cor. 7.6. The hypothesis also is stronger: it requires that f_0 has a whole neighborhood in $\mathcal{P}_d$ formed by polynomials having no indifferent cycle–you could say no parabolic cycle, and that would be an equivalent hypothesis.

(2) In the case $d = 2$, for polynomials of the form $z \mapsto z^2 + c$, the set S is the boundary of the Mandelbrot set. Cor. 8.2 asserts in particular that, for c_0 and c_1 in the same connected component of $\overset{\circ}{M}$, the sets $K(f_{c_0})$ and $K(f_{c_1})$ are homeomorphic, as well as $J(f_{c_0})$ and $J(f_{c_1})$.

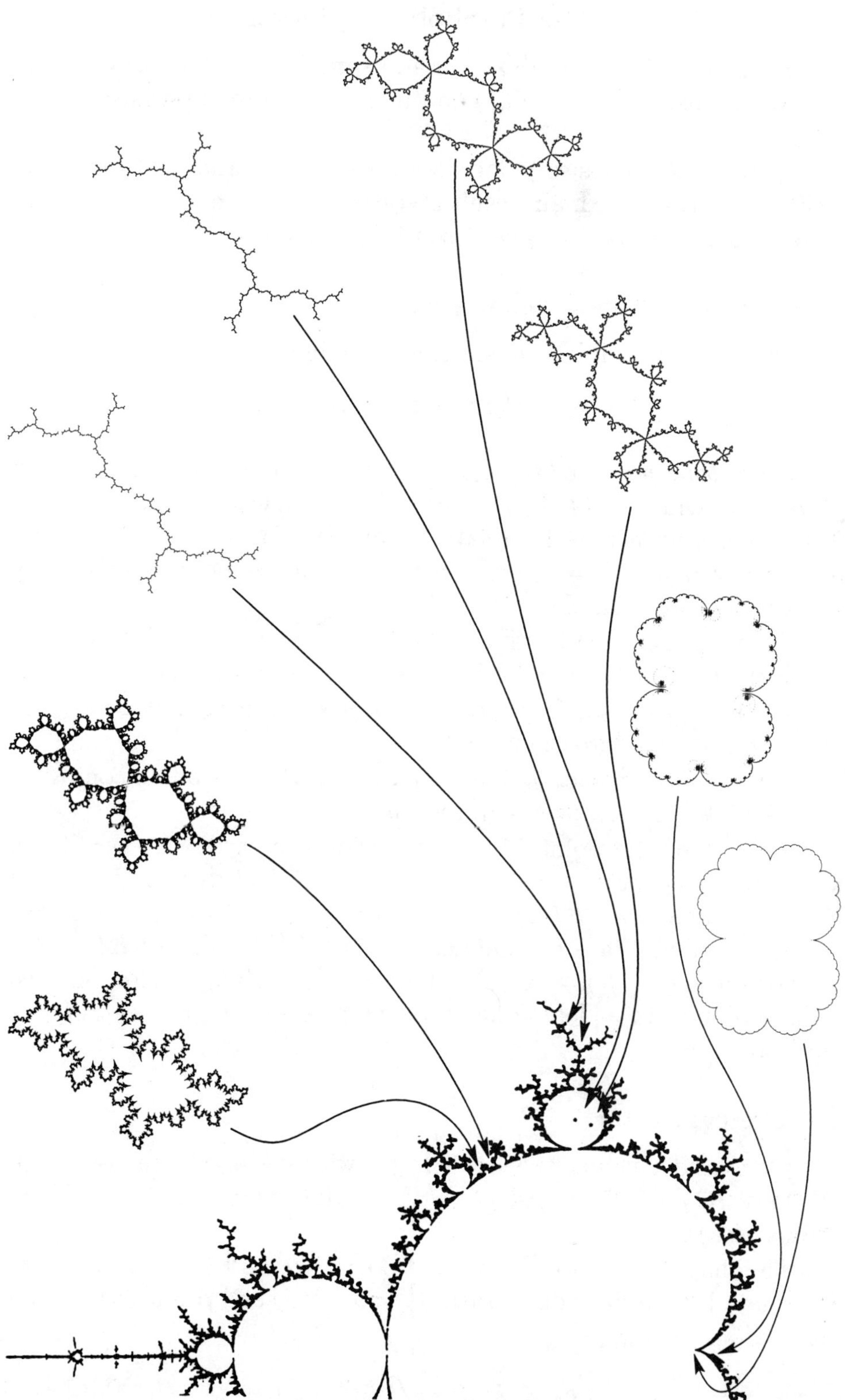

II **The Parabolic Implosion.**

In this part II, we analyze, on an example, the discontinuities at a point f_0 caused in $f \mapsto K(f)$ and in $f \mapsto J(f)$ by a parabolic cycle for f.

§§9 to 13 are expository and descriptive. The paper becomes more technical at §14. The main result is stated in §11 (Th. 11.3). Sections 14 to 21 are dedicated to give essentially the proof of it.

9. **A typical parabolic example.**

The example will be the simplest possible:

$$f_0 \colon z \mapsto z + z^2.$$

A *parabolic point* for a polynomial f of degree d is a point α with $f^k(\alpha) = \alpha$ and $\rho = (f^k)'(\alpha) = e^{2\pi i p/q}$ for some integers $k > 0, q > 0, 0 \le p < q$ with p and q relatively prime. For $f_0 \colon z \mapsto z + z^2$, we have $d = 2, \alpha = 0, k = 1, q = 1, p = 0$. In the sequel we shall study thoroughly this example.

The polynomial f_0 is affine conjugate to $z \mapsto z^2 + 1/4$. The point $1/4$ is the cusp of the big cardioid in the Mandelbrot set

The filled Julia set $K(f_0)$ is the *cauliflower* centered at $\omega = -1/2$. Its boundary is a Jordan curve.

Let us now consider $f_\epsilon \colon z \mapsto z + z^2 + \epsilon$, which is affine conjugate to $z \mapsto z^2 + 1/4 + \epsilon$, for ϵ real, tending to 0.

For $\epsilon < 0$, the point $1/4 + \epsilon$ is inside the cardioid, there is an attracting fixed point, $K(f_\epsilon)$ is still a topological disc, $J(f_\epsilon)$ is still a Jordan curve, and $K(f_\epsilon) \to K(f_0), J(f_\epsilon) \to J(f_0)$ as $|\epsilon| \to 0$. For $\epsilon > 0$, the point $1/4 + \epsilon$ is outside the Mandelbrot set, so $K(f_\epsilon)$ is a Cantor set and $J(f_\epsilon) = K(f_\epsilon)$. Since $K(f_0) \ne J(f_0)$, there must be a discontinuity in at least one of the maps $\epsilon \mapsto K(f_\epsilon)$, $\epsilon \mapsto J(f_\epsilon)$ at the point $\epsilon = 0$.

10. **The eggbeater.**

We consider now $f_\epsilon \colon z \mapsto z + z^2 + \epsilon$ with $\epsilon \in]0, \epsilon_0[$, (say $\epsilon_0 = \frac{1}{25}$). We have $K(f_\epsilon) \cap \mathbf{R} = \emptyset$, since $f_\epsilon(x) > x$ for every real x, so all real points escape.

The map f_ϵ has two fixed points $\alpha = i\sqrt{\epsilon}$ and $\overline{\alpha} = -i\sqrt{\epsilon}$. If we conjugate by the dilation of ratio $\frac{1}{\sqrt{\epsilon}}$, so as to send α and $\overline{\alpha}$ to i and $-i$, the map becomes

$$F_\epsilon \colon Z \to Z + \sqrt{\epsilon}\,(Z^2 + 1),$$

which is the Euler method map with step size $\sqrt{\epsilon}$ for solving the differential equation

$$\dot{Z} = Z^2 + 1.$$

The integral curves of this differential equation are circles, forming a "linear pencil" with limit points i and $-i$ (see Fig.) The orbits of F_ϵ follow approximately these circles. This is what we call the *eggbeater* dynamic.

This description is valid on a fixed compact in the Z-plane, shrinking in the z-plane as $\epsilon \to 0$.

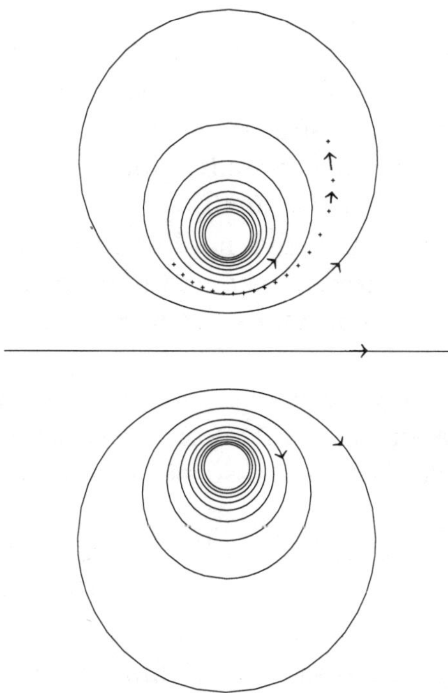

While the points turn on the circles, they also move slightly from α or $\overline{\alpha}$ towards $\mathbf{R}$, since each step is taken on the tangent. The multipliers are $1 \pm 2i\sqrt{\epsilon}$, so α and $\overline{\alpha}$ are slightly repelling, the main effect around them is the rotation.

11. Imaginable scenarios.

What happens to the Cantor set $K(f_\epsilon) = J(f_\epsilon)$ as $\epsilon > 0$ tends to 0? A priori, one could imagine two scenarios. In order to describe them, let us introduce some notation. The triadic Cantor set $C_{1/3}$ is obtained from $I = [0, 1]$ by the following recipe: remove the open middle third, then remove the open middle third from each of the closed intervals remaining, repeat this process and take the intersection of the compact sets obtained at each step. For $\delta > 0$, denote by C_δ the Cantor set obtained by the same recipe, except that at each step you remove from each closed interval of length ℓ a centered open interval of length $\delta \cdot \ell$. When $\delta \to 0$, the Cantor set C_δ tends to I (in terms of the Hausdorff metric).

Scenario 11.1 – For $\epsilon > 0$ small, $J(f_\epsilon) = K(f_\epsilon)$ sits in a Jordan curve close to $J(f_0)$ just like C_δ sits in I for δ small. In other words there is a continuous family Γ_ϵ of Jordan curves with $\Gamma_0 = J(f_0)$, and $J(f_\epsilon)$ is obtained from Γ_ϵ by removing an infinity of very small arcs.

Scenario 11.2 – For $\epsilon > 0$ small, $J(f_\epsilon) = K(f_\epsilon)$ is obtained from $\breve{K}(f_0)$ by making a thin slit along the segment $K(f_0) \cap \mathbf{R}$, and an infinity of other very thin slits which completely disconnect it. So $K(f_\epsilon)$ sits in $K(f_0)$ (or in a closed topological disc close to it) like $C_\delta \times C_\delta$ in the square I^2 for δ small.

The first scenario would give no discontinuity for J, but one for K, with $K(f_0)$ not approximately contained in $K(f_\epsilon) = J(f_\epsilon)$. The second would give no discontinuity for K, but one for J, with $J(f_\epsilon) = K(f_\epsilon)$ not approximately contained in $J(f_0)$ (see plate 3).

Now, what actually occurs is neither of these scenarios. It is much more horrible, or beautiful according to your taste.... let us say more exciting. First there is a discontinuity in both $\epsilon \mapsto J(f_\epsilon)$ and $\epsilon \mapsto K(f_\epsilon)$ at $\epsilon = 0$.

Does this mean that there is a limit L with $J(f_0) \underset{\neq}{\subsetneq} L \underset{\neq}{\subsetneq} K(f_0)$ for $J(f_\epsilon) = K(f_\epsilon)$ as $\epsilon > 0$ tends to 0, or that there is no limit whatsoever?

Well, there *has* to be some kind of limit, for the following reason: For $\epsilon \leq 1$, $K(f_\epsilon)$ is contained in the closed disc $\overline{D}_4$. Now the space $\mathrm{Comp}^*(\overline{D}_4)$ is compact. I did not mention it in §1, but for any compact metric space E the space $\mathrm{Comp}^*(E)$ is compact for the Hausdorff metric. So, for any sequence $\epsilon_n \to 0$ there is a subsequence ϵ_{n_k} such that $K(f_{\epsilon_{n_k}})$ has a limit L. We have $J(f_0) \subset L \subset K(f_0)$ by the semi-continuity properties, and in fact we always have $J(f_0) \underset{\neq}{\subsetneq} L \underset{\neq}{\subsetneq} K(f_0)$. But L does depend on the sequence we started with, and possibly on the subsequence chosen, so $\epsilon \mapsto J(f_\epsilon) = K(f_\epsilon)$ does not have a limit. As we shall see, what occurs is the following:

Theorem 11.3: *One can find a continuous map $\tau\colon {]0, \epsilon_0]} \to \mathbf{T}$ called the phase map, with a lift $\tilde{\tau}\colon {]0, \epsilon_0]} \to \mathbf{R}$ which tends to $-\infty$ as $\epsilon \to 0$ (in particular $\tau(\epsilon)$ does not have a limit), and an injective*

Scenario 1.

Scenario 2.

continuous map $\theta \mapsto L_\theta$ from $\mathbf{T}$ to $\mathrm{Comp}^* K(f_0)$, so that $K(f_{\epsilon_\nu}) \to L_\theta$ for any sequence (ϵ_ν) such that $\epsilon_\nu \to 0$ and $\tau(\epsilon_\nu) \to \theta$.

We have $J(f_0) \underset{\neq}{\subseteq} L_\theta \underset{\neq}{\subseteq} K(f_0)$ for all values of θ.

In the space $\mathrm{Comp}^*(\mathbf{C})$ the points L_θ make a topological circle, the curve $(K(f_\epsilon))_{\epsilon>0}$ spirals asymptotically to this circle as $\epsilon \to 0$. The points $K(f_0)$ and $J(f_0)$ are isolated from this figure.

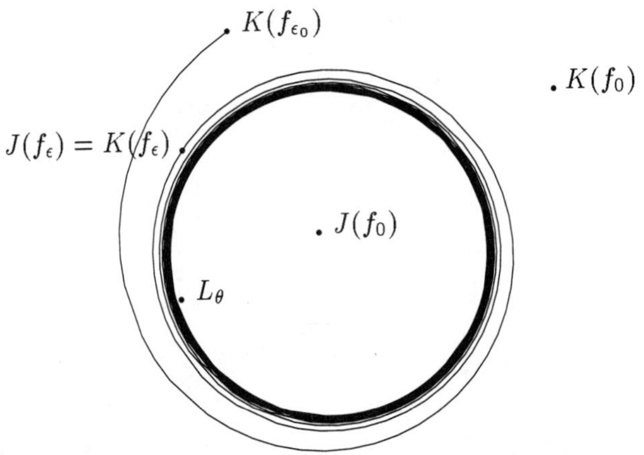

12. The implosion.

This paragraph is just descriptive.

Computer pictures show something very similar to scenario 1, except that there are loops (in interrupted lines) which spiral inward (Plate 2). The diameter of these loops does not tend to 0 when $\epsilon \to 0$. Each (big) loop carries a sequence of pairs of smaller loops. When ϵ moves, the small loops travel along the big loops, and when the phase has made one turn, the sequence has come back to approximately the same position, just shifted, and the pictures are practically indistinguishable. By the same time, the very small loops on the small loops also come back globally to the same position.

Let us follow a point x_ϵ from $\epsilon = 1$ to ϵ approaching 0 according to the maps $\varphi_{1,\epsilon} \cdot J(f_1) \to J(f_\epsilon)$ defined in §8. In other words, let us set

$$x_\epsilon = \varphi_{1,\epsilon}(x_1)$$

with x_1 chosen in $J(f_1)$.

When $\epsilon \to 0$, the point x_ϵ tends to a point $x_0 \in J(f_0)$ and this defines a map $\varphi_{1,0}: J(f_1)$ $\to J(f_0)$. This map is surjective but not injective: points in the backward orbit of $\alpha_0 = 0$ have two preimages, other points have just one. The convergence of $\varphi_{1,\epsilon}$ to $\varphi_{1,0}$ is pointwise, not uniform, even though $\varphi_{1,0}$ is continuous. We know that since

$$\partial(J(f_\epsilon), J(f_0)) \not\to 0.$$

If now we imagine ϵ moving from 0 to 1, points in the backward orbit of $\alpha_0 = 0$ split in two points as soon as ϵ becomes > 0. The other points each have a continuous movement, but there is no uniformity in this continuity. Some points move inwards very fast, so a fixed region in $\overset{\circ}{K}(f_0)$ at distance > 0 from $J(f_0)$ may start being invaded by points of $J(f_\epsilon)$ as soon as ϵ becomes > 0.

I try to express this situation picturesquely by saying that $J(f_0)$ undergoes an implosion, like a T.V. tube full of vacuum.

13. The cause of the discontinuity.

The map $f \mapsto K(f)$ can be factorized as

$$f \mapsto \{f^n\}_{n \in \mathbf{N}} \mapsto K(f).$$

In this section, we try to explain that the cause of the discontinuity lies in the firt map of this factorization, and similarly for $f \mapsto J(f)$.

Fix real points $a < 0$ and $b > 0$ rather small, say $a = -\frac{1}{8}$ and $b = \frac{1}{8}$. For $\epsilon > 0$ we have $f^n(a) \to \infty$ as $n \to \infty$; define the *eggbeater time* $T(\epsilon)$ as the smallest n such that $f_\epsilon^n(a) \geq b$. The point $f_\epsilon^{T(\epsilon)}(a)$ lies in $[b, f_\epsilon(b)]$. So from any sequence of values of ϵ tending to 0 we can extract a sequence (ϵ_n) so that $f_{\epsilon_n}^{T(\epsilon_n)}(a)$ has a limit b^*.

Proposition 13.1: (P. Lavaurs) – *For such a sequence,* $f_{\epsilon_n}^{T(\epsilon_n)}$ *converges uniformly on each compact subset of* $\overset{\circ}{K}(f_0)$ *to a holomorphic map* $g: \overset{\circ}{K}(f_0) \to \mathbf{C}$, *which commutes with* f_0.

This proposition will be proved in §18, after we have introduced coordinates in which the work is particularly easy. These are the Fatou coordinates.

Proposition 13.1 casts some light on the cause of the discontinuity.

Define a *dynamic* on $\mathbf{C}$ as a set $\mathcal{G}$ of maps $g: U \to \mathbf{C}$, U open in $\mathbf{C}$, stable under restriction and composition, i.e. such that

(1) $(g: U \to \mathbf{C}) \in \mathcal{G}$, $U' \subset U \Rightarrow g|U' \in \mathcal{G}$.

(2) $(g_1: U_1 \to \mathbf{C}) \in \mathcal{G}$, $(g_2: U_2 \to \mathbf{C}) \in \mathcal{G}$, $g_1(U_1) \subset U_2 \Rightarrow g_2 \circ g_1 \in \mathcal{G}$.

The dynamic $\mathcal{G}$ is *closed* if, for any sequence $(g_n: U_n \to \mathbf{C})$ in $\mathcal{G}$, if $U_n \to U$ and $g_n \to g$ uniformly on each compact in U, we have $(g: U \to \mathbf{C}) \in \mathcal{G}$.

We say that a sequence $(\mathcal{G}_\nu)$ of dynamics on $\mathbf{C}$ *converges geometrically* to a closed dynamic $\mathcal{G}$ if

(1) For $(g: U \to \mathbf{C}) \in \mathcal{G}$, one can find a sequence $(g_\nu: U_\nu \to \mathbf{C})$ with $g_\nu \in \mathcal{G}_\nu$, $U_\nu \to U$, $g_\nu \to g$ uniformly on each compact in U,

(2) if $\nu_k \to \infty$, $g_{\nu_k}: U_{\nu_k} \to \mathbf{C} \in \mathcal{G}_{\nu_k}$ for each k, $U_{\nu_k} \to U$, $g_{\nu_k} \to g$ uniformly on each compact set in U, then $(g: U \to \mathbf{C}) \in \mathcal{G}$.

The notion of geometric limit is familiar to people working with Kleinian groups.

For $f: \mathbf{C} \to \mathbf{C}$ a polynomial, the *dynamic* $[f]$ *generated* by f is the set of restrictions of iterates of f. Returning to the polynomials $f_\epsilon: z \mapsto z + z^2 + \epsilon$, we have the following: For each ϵ, the dynamic $[f_\epsilon]$ is closed. But $[f_0]$ is *not* the geometric limit of $[f_\epsilon]$ as $\epsilon > 0$ tends to 0. For a sequence ϵ_n as in Prop. 13.1, the map $g: \overset{\circ}{K}(f_0) \to \mathbf{C}$ belongs to the limit of $[f_{\epsilon_n}]$ but not to $[f_0]$.

We don't really have a map $\mathcal{G} \mapsto K(\mathcal{G})$ with good continuity property, but still we can assert that it is the discontinuity in $f \mapsto [f]$ which is reflected in $f \mapsto K(f)$.

14. Fatou coordinates

Let us go back to the map $f_0: z \mapsto z + z^2$. One can find charts in which the expression of f_0 is $z \mapsto z + 1$. More precisely, denote by V^+ and V^- the open discs of diameter $]0, 1/4[$ and $]-1/4, 0[$ respectively.

Lemma 14.1:

(a) The map f_0 is injective on V^+ and V^-.

(b) We have $f_0(V^+) \supset V^+$ and $f_0(V^-) \subset V^-$.

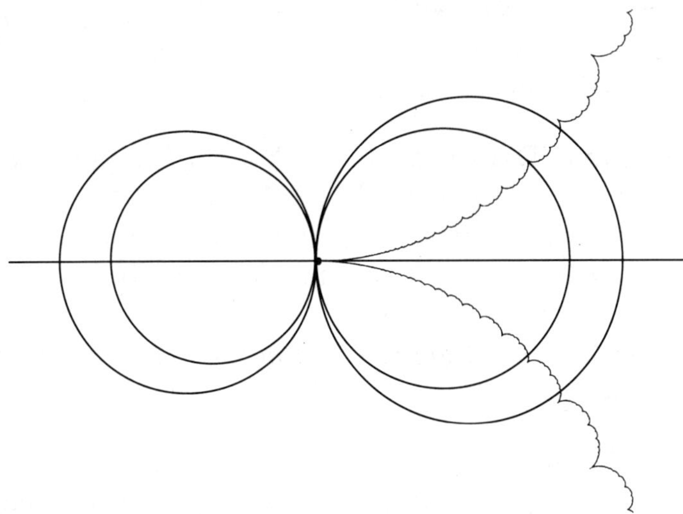

Proof: Let us make the change of variable $Z = \frac{-1}{z}$. The map $f_0 \colon z \mapsto z_1 = z + z^2$ becomes $F \colon Z \mapsto Z_1 = Z + 1 + \frac{1}{Z-1}$. The discs V^+ and V^- become the half planes $U^+ = \{z = x + iy \,|\, x < -4\}$ and $U^- = \{z = x + iy \,|\, x > 4\}$ respectively. On $U^+ \cup U^-$ the map F behaves like the translation $Z \mapsto Z + 1$: the error term $\frac{1}{Z-1}$ is bounded by $\frac{1}{3}$ and its derivative by $\frac{1}{9}$.

So F is injective on U^+ and U^-. Also, $F(U^-) \subset U^-$ and $F(U^+) \supset U^+$. The lemma follows.

q.e.d.

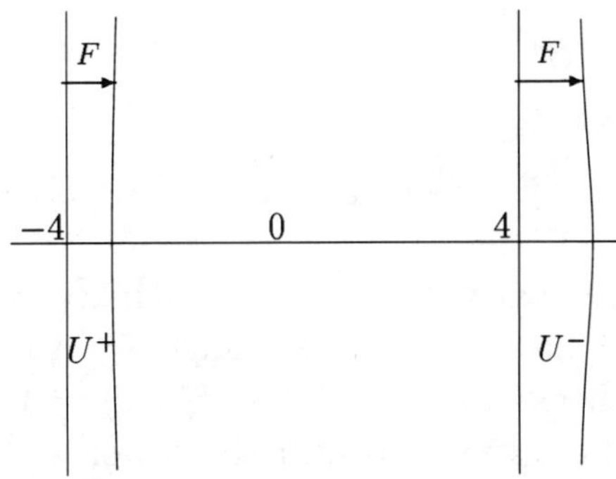

We shall now modify the change of variable $z \mapsto Z$ so as to get rid of the error term. We call a *left region* (resp. a *right region*) a set in $\mathbf{C}$ of the form $\{z = x + iy \mid x < h(y)\}$ (resp. $x > h(y)$), where $h: \mathbf{R} \to \mathbf{R}$ is a continuous function.

Theorem and Definition 14.2: *One can find injective holomorphic maps $\phi^+: V^+ \to \mathbf{C}$ and $\phi^-: V^- \to \mathbf{C}$ such that*

(a) $\phi^+(V^+)$ is a left region and $\phi^-(V^-)$ is a right region;
(b) $\phi^+(f_0(z)) = \phi^+(z) + 1$ if z and $f_0(z)$ are in V^+.
$\phi^-(f_0(z)) = \phi^-(z) + 1$ if $z \in V^-$.
Such maps are called Fatou coordinates. They are unique up to an additive constant.

Sketch of proof. We use the notations of the proof of 14.1. Denote by $X^+ = U^+/F$ the quotient of U^+ by the equivalence identifying Z and $F(Z)$ whenever both are in U^+. Then X^+ is a topological cylinder.

It follows from the Riemann mapping theorem that any Riemann surface homeomorphic to a cylinder is $\mathbf{C}$-analytically isomorphic to $B_h/\mathbf{Z}$ where $B_h = \{z = x + iy \mid |y| < h/2\}$ for some $h \in]0, +\infty]$ (called the modulus), or to $H/\mathbf{Z}$ where H is the upper half plane ("one sided infinite modulus"). Here one can show that the modulus is two-sided infinite, i.e. $X^+ \approx \mathbf{C}/\mathbf{Z}$.

Take an isomorphism $\Psi: X^+ \to \mathbf{C}/\mathbf{Z}$ and denote by χ_F the quotient map $U^+ \to X^+$. Since U^+ is simply connected, the map $\Psi \circ \chi_F: U^+ \to \mathbf{C}/\mathbf{Z}$ can be lifted to a map $\phi_u^+: U^+ \to \mathbf{C}$. Composing with the change of variable $z \mapsto Z = \frac{-1}{z}$, we get a map $\phi^+: V^+ \to \mathbf{C}$. By construction, we have a commutative diagram

$$
\begin{array}{ccccc}
V^+ & \overset{z \mapsto Z}{\underset{\approx}{\rightrightarrows}} & U^+ & \overset{\phi_u^+}{\longrightarrow} & \mathbf{C} \\[2em]
\downarrow \chi_f & & \downarrow \chi_F & & \downarrow \chi_{\mathbf{z}} \\[2em]
V^+/f & \overset{\approx}{\longrightarrow} & U^+/F & \overset{\Psi}{\longrightarrow} & \mathbf{C}/\mathbf{Z}
\end{array}
$$

If Z and Z_1 are two points in U^+ with $Z_1 = F(Z)$, the points $\phi_u^+(Z)$ and $\phi_u^+(Z_1)$ have the same image by $\chi_{\mathbf{z}}$, i.e. differ by an integer. One can check that $\phi_u^+(Z_1) - \phi_u^+(Z) = 1$ (provided ψ maps the upper end to the upper end and the lower end to the lower end) and that $\phi^+: V \to \mathbf{C}$ defined by $\phi^+(z) = \phi_u^+(Z)$ has the desired properties.

One constructs similarly $\phi^-\colon V^- \to \mathbf{C}$.

It is easy to see that Fatou coordinates are necessarily obtained by the above construction. There are some choices in this construction:

- choice of Ψ (but unless you allow to interchange $+i\infty$ and $-i\infty$, the isomorphism Ψ can only be changed by composing with a translation in the group $\mathbf{C}/\mathbf{Z}$).
- choice of the lift ϕ of $\Psi \circ \chi$ (you can change it only by an integer translation).

These choices affect the Fatou coordinates only by adding a constant.

q.e.d.

Remarks. One can choose the Fatou coordinates so that they are real on $\mathbf{R} \cap V^+$ and $\mathbf{R} \cap V^-$.

15. Extension of the Fatou coordinates

Proposition 15.1: *(a) The attracting Fatou coordinate $\phi^-\colon V^- \to$ $\mathbf{C}$ can be extended into a holomorphic map $\hat\phi^-\colon \overset{\circ}{K}(f_0) \to \mathbf{C}$ satisfying*

$$\hat\phi^-(f_0(z)) = \hat\phi^-(z) + 1$$

for all $z \in \overset{\circ}{K}(f_0)$.

(b) the inverse $\psi^+\colon \phi^+(V^+) \to V^+$ of the repelling Fatou coordinate can be extended into a map $\hat\psi^+\colon \mathbf{C} \to \mathbf{C}$ satisfying $\psi^+(z+1) = f_0(\psi^+(z))$ for all $z \in \mathbf{C}$.

Proof: We use the notations of the proof of 14.1. For $Z \in U^-$, we have $|F^n(Z)| \to \infty$ and $\operatorname{Arg} F^n(Z) \to 0$ as $n \to \infty$, so for $z \in V^-$ the point $f_0^n(z)$ tends to 0 tangentially to $\mathbf{R}_-$. At the parabolic point 0, the unique attracting axis is $\mathbf{R}_-$ and $\mathbf{R}_+$ is the repelling axis. The open disc V^- is contained in the basin of 0, and this basin is $\overset{\circ}{K}(f_0)$ since $\overset{\circ}{K}(f_0)$ is connected.

(a): For $z \in \overset{\circ}{K}(f_0)$, we have $f_0^n(z) \in V^-$ for n large enough. We then set $\hat\phi^-(z) = \phi^-(f_0^n(z)) - n$, noting that this does not depend on the choice of n.

(b) For $z \in \mathbf{C}$, we have $z - n \in U^+$ for n great enough. We then set $\hat\psi^+(z) = f_0^n(\psi^+(z-n))$, noting that this does not depend on the choice of n.

One checks easily that the maps $\hat{\phi}^-$ and $\hat{\psi}^+$ satisfy the requirements.

q.e.d.

We call $\hat{\phi}^-$ the extended Fatou coordinate and $\hat{\psi}^+$ the extended Fatou parametrization. Set $\omega = -\frac{1}{2}$; this point is the critical point of f_0, and it is a center of symmetry for $K(f_0)$. Let us adjust the attracting Fatou coordinate $\hat{\phi}_0^-$ by the condition $\hat{\phi}_0^-(\omega) = 0$.

Then we have:

Proposition 15.2: *The map $\hat{\phi}^-: \overset{\circ}{K}(f_0) \to \mathbf{C}$ is a ramified covering with infinitely many sheets, ramified only over the non-positive integers. The critical points of $\hat{\phi}^-$ are the precritical points of f_0 i.e. points in $\bigcup f_0^{-n}(\omega)$.*

We don't have such a nice description for the map $\hat{\psi}^+$.

Rather than proving Proposition 15.2, we shall give, without proof, a more detailed description of the map $\hat{\phi}^-$.

Let us color in yellow the open upper half of V^-, in brown the lower upper half and in purple the diameter $V^- \cap \mathbf{R}$. Note that $f_0(z)$ has the same color than z. Extend this coloring to $\overset{\circ}{K}(f_0)$ by coloring z in the same color than $f_0^n(z)$ for n large. We obtain in this way what we call the *parabolic chessboard*: the purple part is $S = \bigcup f_o^{-n}]-1, 0[$, the boxes are the connected components of $\overset{\circ}{K}(f_0) - S$. The two *main boxes* are the boxes B' and B'' which contain the upper and lower half of V^-.

The map f_0 induces a biholomorphic homeomorphism of B' onto itself, and $\hat{\phi}^-$ induces a biholomorphic homeomorphism of B' onto the upper half plane H, thus realizing the Riemann mapping. Since $\hat{\phi}^-$ conjugates f_0 to the translation $\zeta \mapsto \zeta + 1$ we can say that $f_0: B' \to B'$ is a parabolic map (maybe this is where the name of parabolic points comes from).

The situation with B'' is symmetric.

The corners of the chess board are the precritical points, i.e., the points in $\bigcup_{n \geq 0} f^{-n}(\omega)$. The map $\hat{\phi}^-$ sends them to nonpositive integers. For each box B, there is an integer $n \geq 0$ such that f_0^n maps B biholomorphically onto B' or B''.

16. Persistence of the Fatou coordinate for the perturbed map.

Let us now consider $f_\epsilon : z \mapsto z + z^2 + \epsilon$ for $\epsilon > 0$ small (say $\epsilon \in]0, \epsilon_0], \epsilon_0 = \frac{1}{25}$). Take for V_ϵ^+ (resp. V_ϵ^-) the disc bounded by the circle through the fixed points $\alpha = i\sqrt{\epsilon}$, $\overline{\alpha} = -i\sqrt{\epsilon}$ and the point $1/4$ (resp. $-1/4$).

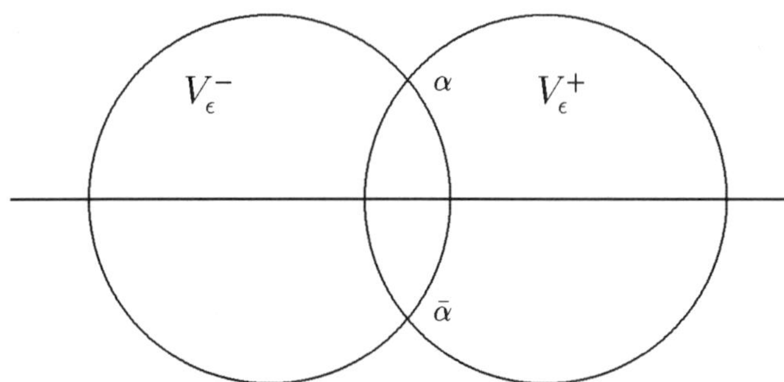

We cannot expect any longer to get a map $\phi^- : V_\epsilon^- \to \mathbf{C}$ conjugating f to $\zeta \mapsto \zeta + 1$ and mapping V_ϵ^- onto a right region, since the orbit of a point in V_ϵ^- always leaves V_ϵ^- in a finite time.

Proposition and Definition 16.1: *One can find injective holomorphic maps $\phi_\epsilon^+ : V_\epsilon^+ \to \mathbf{C}$, $\phi_\epsilon^- : V_\epsilon^- \to \mathbf{C}$ such that:*

(a) $\phi_\epsilon^+(V_\epsilon^+)$ is a region of the form $\{\zeta = \xi + i\eta | h_1^+(\eta) < \xi < h^+(\eta)\}$ with $h^+, h_1^+ : \mathbf{R} \to \mathbf{R}$ continuous, $h^+ - h_1^+ > 1$, and similarly for ϕ_ϵ^-;

(b) $\phi_\epsilon^+(f_\epsilon(z)) = \phi_\epsilon^+(z) + 1$ whenever z and $f_\epsilon(z)$ are both in V_ϵ^+ and similarly for ϕ_ϵ^-.

We still call such maps Fatou coordinates. They are unique up to additive constants.

Sketch of proof: We shall adapt the proof of Prop. 14.2 The change of variable which is convenient here is

$$z \mapsto Z = \frac{1}{2\alpha} \, \mathrm{Log} \frac{z - \alpha}{z - \overline{\alpha}}$$

We choose on V_ϵ^- (resp. V_ϵ^+) the branch of Log which is close to 0 for $z = \frac{-1}{4}$ (resp. $z = \frac{1}{4}$).

This transforms V_ϵ^- and V_ϵ^+ into vertical strips U_ϵ^- and U_ϵ^+. The map $f_\epsilon \colon z \mapsto z_1$ becomes a map $F_\epsilon \colon Z \mapsto Z_1$ defined by $Z_1 = Z + \frac{1}{\sqrt{\epsilon}} \arctan \frac{\sqrt{\epsilon}}{1+z}$. For $|z| \le \frac{1}{4}$, $\frac{1}{\sqrt{\epsilon}} \arctan \frac{\sqrt{\epsilon}}{1+z}$ is close enough to $\frac{1}{1+z}$ to make sure that its real part is > 0. The quotients $X_\epsilon^+ = U_\epsilon^+/F_\epsilon$ and $X_\epsilon^- = U_\epsilon^-/F_\epsilon$ are **C**-analytic annuli, i.e. Riemann surfaces homeomorphic to cylinders, and one can prove that their modulus is two-sided infinite. The proof then continues like for 14.2.

q.e.d.

Remark: With the convention that we have taken for the Log, the map

$$z \mapsto Z = \frac{-i}{2\sqrt{\epsilon}} \operatorname{Log} \frac{z - i\sqrt{\epsilon}}{z + i\sqrt{\epsilon}}$$

tends to the map $z \mapsto Z = \frac{-1}{z}$ considered in §14, when $\epsilon \to 0$. In order to obtain this advantage, we have sacrificed another one. On $V_\epsilon^- \cap V_\epsilon^+$, we have two values for Z which differ by $\frac{\pi}{\sqrt{\epsilon}}$. With another choice, we could make them match and get one Fatou coordinate on $V_\epsilon^- \cup V_\epsilon^+$.

17. The phase

Do the Fatou coordinates ϕ_ϵ^- and ϕ_ϵ^+ tend to ϕ_0^- and ϕ_0^+ when $\epsilon > 0$ tends to 0?

This question has to be made more precise for two reasons:

(1) The domain of definition depends on ϵ. This is not a serious issue since $\mathcal{V}^- = \{(\epsilon, z) | z \in V_\epsilon^-\}$ and $\mathcal{V}^+ = \{(\epsilon, z) | z \in V_\epsilon^+\}$ are open in $[0, \epsilon_0] \times \mathbf{C}$.

(2) The maps ϕ_ϵ^- and ϕ_ϵ^+ are only well defined up to a constant. For this, let us fix base points in $V_\epsilon^\pm \cap \mathbf{R}$ independent of ϵ, say $a = \frac{-1}{8}$ and $b = \frac{1}{8}$. For each $\epsilon \in [0, \epsilon_0]$ let us normalize ϕ_ϵ^- and ϕ_ϵ^+ by $\phi_\epsilon^-(a) = \phi_\epsilon^+(b) = 0$.

Proposition 17.1 *With all of these conventions, $(\epsilon, z) \mapsto \phi_\epsilon^-(z)$ and $(\epsilon, z) \mapsto \phi_\epsilon^+(z)$ are continuous on $\mathcal{V}^-$ and $\mathcal{V}^+$ respectively.*

The proof is not really hard, but it involves some complex analysis. We don't give it here.

For $\epsilon > 0$, the map ϕ_ϵ^- and ϕ_ϵ^+ both induce Fatou coordinates on $V_\epsilon^- \cap V_\epsilon^+$, so they differ by a constant. We call this constant the *lifted phase* $\tilde{\tau}(\epsilon)$, thus setting

$$\tilde{\tau}(\epsilon) = \phi_\epsilon^+(z) - \phi_\epsilon^-(z) \ \text{ for } z \in V_\epsilon^- \cap V_\epsilon^+.$$

Remark 17.2: The uniqueness of the Fatou coordinate on $V_\epsilon^- \cap V_\epsilon^+$ up to an additive constant relies on the fact that $V_\epsilon^- \cap V_\epsilon^+ / f_\epsilon$ is a cylinder. One can check that each orbit in V_ϵ^- enters $V_\epsilon^- \cap V_\epsilon^+$ (it cannot jump over it), so $V_\epsilon^- \cap V_\epsilon^+ / f_\epsilon$ can be identified with $X_\epsilon^- = V_\epsilon^- / f_\epsilon$.

The lifted phase $\tilde{\tau}(\epsilon)$ is related to the eggbeater time T_ϵ (smallest n such that $f_\epsilon^n(a) \geq b$) by:

Proposition 17.3 *We have*

$$-T_\epsilon \leq \tilde{\tau}(\epsilon) < -T_\epsilon + 1.$$

Proof: Let us set $a' = f_\epsilon^{T_\epsilon}(a)$, and $a'' = f_\epsilon^m(a)$ with m chosen so that $a'' \in V_\epsilon^- \cap V_\epsilon^+$.

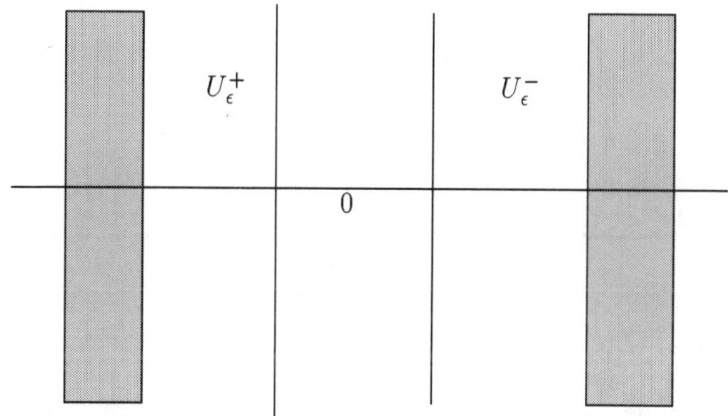

We have $\phi_\epsilon^-(a'') = m$ and $\phi_\epsilon^+(a'') = \phi_\epsilon^+(a') - (T_\epsilon - m)$, so $\tilde{\tau}(\epsilon) = \phi_\epsilon^+(a'') - \phi_\epsilon^-(a'') = -T_\epsilon + \phi_\epsilon^+(a')$. Now $a' \in [b, f_\epsilon(b)[$ by definition of T_ϵ, and ϕ_ϵ^+ maps $[b, f_\epsilon(b)[$ onto $[0, 1[$, so $0 \leq \phi_\epsilon^+(a') < 1$.

q.e.d.

Corollary 17.4 *The lifted phase* $\tilde{\tau}(\epsilon)$ *tend to* $-\infty$ *when* $\epsilon > 0$ *tends to 0.*

Indeed $T_\epsilon \to \infty$, since $f_\epsilon^n(a) \to f_0^n(a) \in [a, 0[$ when $\epsilon \to 0$ with n fixed.

Remark 17.5. One can show that $\tilde{\tau}(\epsilon) = \frac{-\pi}{\sqrt{\epsilon}} + O(1)$.

We denote by $\tau(\epsilon)$ the class of $\tilde{\tau}(\epsilon)$ in $\mathbf{T} = \mathbf{R}/\mathbf{Z}$, and we call it the *phase*.

If $\tau(\epsilon) = 0$, the orbit of a contains that of b. In general, $\tau(\epsilon)$ tells how the orbit of a is situated with respect to the orbit of b (on the right of b). As a consequence of Cor. 17.4, we get

Corollary 17.6 *The phase* $\tau(\epsilon)$ *does not have a limit when* $\epsilon \to 0$.

For $\sigma \in \mathbf{C}$, let us denote by T_σ the translation $Z \mapsto Z + \sigma$.

Proposition 17.7. *For* $\epsilon > 0$ *and* $n \in \mathbf{N}$, *the map*

$$(\phi_\epsilon^+)^{-1} \circ T_{n+\tilde{\tau}(\epsilon)} \circ \phi_\epsilon^-$$

coincides with f_ϵ^n *wherever it is defined.*

Proof: Let $\zeta \in \Omega_\epsilon^- = \phi_\epsilon^-(V_\epsilon^-)$ be such that $\zeta + n + \tilde{\tau}(\epsilon) \in \Omega_\epsilon^+$. Let m be the largest integer $\leq n$ such that $\zeta + m \in \Omega_\epsilon^-$. Then $\zeta + m + \tilde{\tau}(\epsilon) \in \Omega_\epsilon^+$, $\zeta + i \in \Omega_\epsilon^-$ for $0 \leq i \leq m$, $\zeta + j + \tilde{\tau}(\epsilon) \in \Omega_\epsilon^+$ for $m \leq j \leq n$. If $\zeta = \phi_\epsilon^-(z)$, one has $f^m(z) = (\phi_\epsilon^-)^{-1}(\zeta + m) = (\phi_\epsilon^+)^{-1}(\zeta + m + \tilde{\tau}(\epsilon))$ and $f^n(z) = f^{n-m}(f^m(z)) = (\Phi_\epsilon^+)^{-1}(\zeta + n + \tilde{\tau}(\epsilon))$.

q.e.d.

Corollary 17.8 *Take* $x \in V_\epsilon^-$ *and* $n \in \mathbf{N}$ *such that* $f_\epsilon^n(x) \in V_\epsilon^+$ *and* $f_\epsilon^i(x) \in V_\epsilon^- \cup V_\epsilon^+$ *for* $0 \leq i \leq n$. *Then the expression of* f_ϵ^n *on a neighborhood of* x, *in the charts* ϕ_ϵ^- *and* ϕ_ϵ^+ *is* $\zeta \mapsto \zeta + n + \tilde{\tau}(\epsilon)$.

18. Lavaurs maps

We are now in position to define the map g which occurs in Prop. 13.1

Definition 18.1: *A Lavaurs map for* f_0 *is a map* $\overset{\circ}{K}(f_0) \to \mathbf{C}$ *of the form* $g_\sigma = \hat{\psi}^+ \circ T_\sigma \circ \hat{\phi}^-$ *where* $\hat{\phi}^-$ *and* $\hat{\psi}^+$ *are extended Fatou coordinate and parametrization of* f_0, $\sigma \in \mathbf{C}$ *and* T_σ *is the translation* $\zeta \mapsto \zeta + \sigma$.

We have $g_\sigma \circ f_0 = f_0 \circ g_\sigma = g_{\sigma+1}$.

The set of Lavaurs maps does not depend on the choice of the Fatou coordinates, but the correspondence $\sigma \mapsto g_\sigma$ does. The following proposition is a refinement of Prop. 13.1:

Proposition 18.2: *Let* ϵ_ν *be a sequence of positive numbers tending to 0 and* (n_ν) *a sequence of integers tending to* $+\infty$. *Suppose that*

$n_\nu + \tilde{\tau}(\epsilon_\nu)$ *tends to a limit* $\sigma \in \mathbf{R}$. *Then* $f_{\epsilon_\nu}^{n_\nu}$ *tends to the Lavaurs map* g_σ *uniformly on each compact subset of* $\overset{\circ}{K}(f_0)$.

Proof: Suppose first $x \in V_0^-$ and σ such that $\phi_0^-(x) + \sigma \in U_0^+ = \phi_0^+(V_0^+)$. Then one can find a neighborhood Λ of x such that

$$(\phi_{\epsilon_\nu}^+)^{-1} \circ T_{h_\nu} + \tilde{\tau}(\epsilon_\nu) \circ \phi_{\epsilon_\nu}^-$$

is defined on Λ for ν large enough, and converges uniformly on Λ to g_σ. On Λ, we have

$$(\phi_{\epsilon_\nu}^+)^{-1} \circ T_{n_\nu + \tilde{\tau}(\epsilon_\nu)} \circ \phi_{\epsilon_\nu}^- = f_{\epsilon_\nu}^{n_\nu}, \text{ so } f_{\epsilon_\nu}^{n_\nu} \to g_\sigma$$

uniformly on Λ.

Let now x be an arbitrary point in $\overset{\circ}{K}(f)$ and $\sigma \in \mathbf{R}$ be arbitrary. One can find $m_1 \in \mathbf{N}$ such that $x' = f_0^{m_1}(x) \in V_0^-$, and $\sigma' = \sigma - m_1 - m_2$ with $m_2 \in \mathbf{N}$ such that $\phi_\epsilon^-(x') + \sigma'' \in U_0^+$.

Then, setting $n_\nu' = n_\nu - m_1 - m_2$, the map $f_{\epsilon_\nu}^{n_\nu}$ tends to $g_{\sigma'}$, uniformly on a neighborhood of x', so $f_{\epsilon_\nu}^{n_\nu} \to f_0^{m_2} \circ g_{\sigma'}, \circ f_0^{m_1} = g_\sigma$ uniformly on a neighborhood of x.

q.e.d.

19. The sets $K(f_0, g_\sigma)$ and $J(f_0, g_\sigma)$

In this section, we define the sets which will occur as possible limits of f_ϵ when $\epsilon \to 0$.

Consider f_0 and a Lavaurs map g_σ. We first define points which escape by (f_0, g_σ).

A point in $\mathbf{C} - K(f_0)$ escapes by f_0, and we say that it escapes by (f_0, g_σ). A point in $J(f_0)$ does not escape by f_0, and we cannot apply g_σ to it, nor to any of its iterated images by f_0: then it does not escape by (f_0, g_σ). To a point z in $\overset{\circ}{K}(f_0)$ we can apply g_σ. If $g_\sigma(z) \in \mathbf{C} - K(f_0)$, it then escapes by f_0, and we say that z escapes by (f_0, g_σ). If $g_\sigma(z) \in J(f_0)$, it is trapped there, so z does not escape. If $g_\sigma(z) \in \overset{\circ}{K}(f_0)$, we can apply g_σ again and it gets another chance. Applying f_0 first does not change anything since f_0 commutes with g_σ, and $K(f_0)$ is fully invariant by f_0.

Finally we define points which escape by (f_0, g_σ) as points z for which $(\exists m \in \mathbf{N})$ such that $g_\sigma^m(z) \in \mathbf{C} - K(f_0)$. This means: there exists $m \in \mathbf{N}$ such that $g_\sigma^m(z)$ is defined and is outside $K(f_0)$. The case $m = 0$ is included, so points in $\mathbf{C} - K(f_0)$ escape. We denote by $K(f_0, g_\sigma)$ the set of points in $\mathbf{C}$ which do not escape by (f_0, g_σ).

We define $J(f_0, g_\sigma)$ as the closure of

$$\{z | (\exists m \in \mathbf{N})\ g_\sigma^m(z) \in J(f_0)\}$$

Proposition 19.1 *The Set $J(f_0, g_\sigma)$ is the boundary of $K(f_0, g_\sigma)$.*

Proof: If $g_\sigma^m(z) \in J(f_0)$, then z cannot escape, so $z \in K(f_0, g_\sigma)$, but arbitrarily close to z one can find z' such that $g_\sigma^m(z') \in \mathbf{C} - K(f_0)$ because g_σ^m is open, so $z \in \mathbf{C} - K(f_0, g_\sigma)$. Therefore $J(f_0, g_\sigma) \subset \partial K(f_0, g_\sigma)$.

Conversely, let z be a point in $\partial(K(f_0, g_\sigma))$, and choose $\delta > 0$. Take z' with $|z' - z| < \delta$ and m such that $g_\sigma^m(z') \in \mathbf{C} - K(f_0)$. Either $g_\sigma^m(z)$ is defined, or $(\exists m' < m)$ such that $g_\sigma^{m'}(z) \in J(f_0)$. In the first case, $g_\sigma^m(z) \in K(f_0)$, and $(\exists z'' \in [z, z'])$ such that $g_\sigma^m(z'') \in J(f_0)$. In both cases $d(z, J(f_0, g_\sigma)) < \delta$. Since this holds for any $\delta > 0$, we have $z \in J(f_0, g_\sigma)$.

q.e.d.

We now restrict for a while our attention to real values of σ.

We have $g_{\sigma+1} = g_\sigma \circ f_0 = f_0 \circ g_\sigma$. It follows that $K(f_0, g_\sigma)$ and $J(f_0, g_\sigma)$ only depends on the class θ of σ in $\mathbf{T} = \mathbf{R}/\mathbf{Z}$. We denote them by $K(f_0, \theta)$ and $J(f_0, \theta)$.

The semi-continuity properties extend to this context: The following proposition generalizes Thm. 5.1.

Proposition 19.2. *Let (ϵ_ν) be a sequence tending to 0 with $\tau(\epsilon_\nu) \to \theta \in \mathbf{T}$. Then*
(a) $\partial(K(f_{\epsilon_\nu}),\ K(f_0, \theta)) \to 0$.
(b) $\partial(J(f_0, \theta),\ J(f_{\epsilon_\nu})) \to 0$.

We shall see in §21 that for $\sigma \in \mathbf{R}$ we actually have $K(f_0, g_\sigma) = J(f_0, g_\sigma)$, so in this case $K(f_{\epsilon_\nu})$ converges to $K(f_0, \theta)$.

Proof: (a) If z escapes by (f_0, g_σ), then any point z' sufficiently close to z escapes by (f_0, g_σ), and also by f_{ϵ_ν} for ν large enough by Prop. 18.2. So the set

$$\mathcal{X} = \bigcup(\{\epsilon_\nu\} \times K(f_{\epsilon_\nu})) \cup (\{0\} \times K(f_0, g_\sigma))$$

is closed in

$$(\{\epsilon_\nu\}_{\nu \in \mathbf{N}} \cup \{0\}) \times \overline{D}_4),$$

and thus is compact. We can then apply Prop. 2.1.

(b) Let $\sigma \in \mathbf{R}$ be a representative of θ, and define $n_\nu \in \mathbf{N}$ for each ν so that $\tilde{\tau}(\epsilon_\nu) + n_\nu \to \sigma$.

Let x_0 be a repelling periodic point for f_0. By the implicit function theorem, we can find an analytic map $\epsilon \mapsto x_\epsilon$ giving x_0 and such that x_ϵ is periodic with the same period for f_ϵ. Points z for which $(\exists m, l)$ such that $f_0^l g_\sigma^m(z) = x_0$, are dense in $J(f_0, g_\sigma)$. For such a point z, we have $g_\sigma^i(z) \in \overset{\circ}{K}(f_0)$ for $0 \le i < m$, and $f_{\epsilon_\nu}^{mn_\nu} \to g_\sigma^m$ uniformly on a neighborhood of z. So one can find a sequence (z_ν) tending to z so that $f^{mn_\nu+l}(z_\nu) = x_{\epsilon_\nu}$. Therefore $d(z, J(f_{\epsilon_\nu})) \to 0$.

By an argument similar to the proof of Thm. 5.1(b), one shows that $\partial(J(f_0, \theta), J(f_{\epsilon_\nu})) \to 0$

q.e.d.

Proposition 19.2 was stated for real values of ϵ, because $\tau(\epsilon)$ has been defined only for $\epsilon \in]0, \epsilon_0]$ up to now. But this definition can be extended to complex values of ϵ lying in some sector (see §22), and in this context Prop. 19.2 extends without change.

20. The case where the critical point escapes.

Denote by ω the critical point of f_0, i.e. $\omega = \frac{-1}{2}$. Take $\sigma \in \mathbf{C}$.

Theorem 20.1 (P. Lavaurs). *If $\omega \notin K(f_0, g_\sigma)$, the set $K(f_0, g_\sigma)$ has empty interior. In other words $K(f_0, g_\sigma) = J(f_0, g_\sigma)$.*

This result is pretty hard to prove, and we shall only indicate the steps of the proof. All results in this section are from Lavaur's thesis.

We say that a point z is *periodic for* (f_0, g_σ) if $(\exists m, l)$, $m + l > 0$ so that $f_0^l g_\sigma^m(z) = z$ (by this we mean $f_0^l g_\sigma^m$ is defined and equal to z). The multiplier ρ of z is then $(f_0^l g_\sigma^m)'(z)$. We say that z is repelling if $|\rho| > 1$, etc.

Points which are periodic for f_0 are periodic for (f_0, g_σ): take $m = 0$.

A periodic point for (f_0, g_σ) is also periodic for $(f_0, g_{\sigma-1})$ since $g_\sigma^m = f_0^m \circ g_{\sigma-1}^m$. We say that z is a *generalized periodic point* for (f_0, g_σ) if it is periodic for $(f_0, g_{\sigma-k})$ for some $k \in \mathbf{N}$.

We also have to take into account the two *virtual multipliers* (or *renormalized multipliers*) of the fixed point α_0 of f_0. They are defined in the following way. The map g_σ induces an isomorphism

$$E_\sigma \colon X_0^- = \overset{\circ}{K}(f_0)/f_0 \xrightarrow{\approx} X_0^+ = V^+/f_0.$$

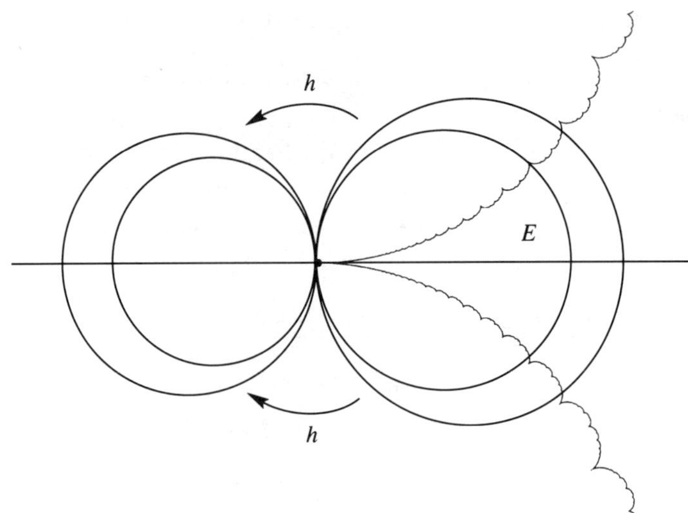

On the other hand a point $\zeta \in X_0^+$ which is close to one of the ends of the cylinder is represented by points in $V^+ \cap \overset{\circ}{K}(f_0)$ which are in the same orbit, so they define a point in the quotient $\overset{\circ}{K}(f_0)/f_0 = X_0^-$. We define in this way the *horns map* $h: X_0^+ - A \longrightarrow X_0^-$, where A is a compact annulus. Composing, we get a map $E_\sigma \circ h: X_0^+ - A \to X_0^+$. We compactify X_0^+ into the Riemann sphere $\overline{\mathbf{C}}$, sending the two ends to 0 and ∞. We get a map

$$L: \overline{\mathbf{C}} - A \to \overline{\mathbf{C}}$$

with fixed points at 0 and ∞, which we can call the *renormalization* of (f_0, g_σ). The multipliers of L at 0 and ∞ are the virtual multipliers of (f_0, g_σ) at α_0. We say that α_0 is virtually attracting on one side and virtually repelling on the other if one of the virtual multipliers has modulus > 1 and the other < 1, etc.

If α_0 is virtually attracting, parabolic or Siegel on one side, one can define its basin or Siegel domain on that side.

Theorems 4.2 and 4.3 of Sullivan and Fatou can be extended to this content:

Proposition 20.2 *Each connected component U of $\overset{\circ}{K}(f_0, g_\sigma)$ is preperiodic in the generalized sense, i.e. one can find l, m, k such that $f_0^l g_{\sigma-k}^m(U)$ is $(f_0, g_{\sigma-k})$ periodic.*

Proposition 20.3 *Each connected component of $\overset{\circ}{K}(f_0, g_\sigma)$ that is periodic is of one of the following types:*

(i) the attracting basin, parabolic basin or Siegel disk of a (f_0, g_σ)-periodic point other than the parabolic fixed point $\alpha_0 = 0$.

(ii) the attracting basin, parabolic basin or Siegel domain of α_0 on one side.

So in order to prove Thm. 20.1, it suffices to prove:

Proposition 20.4 If $\omega \notin K(f_0, g_\sigma)$, all periodic points for (f_0, g_σ) are repelling, except for the parabolic fixed point α_0, which is virtually repelling on both sides.

21. The case σ real.

Proposition 21.1 *For $\sigma \in \mathbf{R}$, we have $K(f_0, g_\sigma) = J(f_0, g_\sigma)$.*

Proof: We have $\hat{\phi}^-(\mathbf{R} \cap \overset{\circ}{K}(f_0)) = \mathbf{R}$, $T_\sigma(\mathbf{R}) = \mathbf{R}$ and $\hat{\psi}^+(\mathbf{R}) = \mathbf{R}^*_+ =]0, \infty[$. So $g_\sigma(\omega) \in \mathbf{R}^*_+ \subset \mathbf{C} - K(f_0)$, thus $\omega \notin K(f_0, g_\sigma)$ and we can apply Thm. 20.1.

q.e.d.

Corollary 21.2 *Let (ϵ_ν) be a sequence in $]0, \epsilon_0]$ tending to 0 with $\tau(\epsilon_\nu) \to \theta \in \mathbf{T}$. Then*

$$d_H(K(f_{\epsilon_\nu}), \ K(f_0, \theta)) \to 0.$$

Proof: This is just Prop. 19.2, considering that we have $K(f_{\epsilon_\nu}) = J(f_{\epsilon_\nu})$ because $c_{\epsilon_\nu} = \frac{1}{u} + \epsilon_\nu$ is outside the Mandelbrot set, and $K(f_0, \theta) = J(f_0, \theta)$ by Prop. 21.1.

q.e.d.

Note that this is a refinement of Thm. 11.3.

22. What for ϵ non real?

Let now ϵ range in the sector S defined by $|\epsilon| \le r_0 \le \frac{1}{25}$, $|\mathrm{Arg}\epsilon| \le \frac{\pi}{4}$, i.e. $\mathrm{Re}(\epsilon) \ge |\mathrm{Im}(\epsilon)|$. The map $f_\epsilon \colon z \mapsto z + z^2 + \epsilon$ has two fixed points $\alpha' = i\sqrt{\epsilon}$ and $\alpha'' = -i\sqrt{\epsilon}$. Between them, we still have an eggbeater dynamic.

One can define V_ϵ^+ and V_ϵ^- for $\epsilon \in S^* = S - \{0\}$ in such a way that

(1) $V_\epsilon^- \to V_0^- = V^-$ in the following sense: $\overline{V_\epsilon^-} \to \overline{V_0^-}$ and $\partial V_\epsilon^- \to \partial V_0^-$ for the Hausdorff distance which implies for any compact $L \subset V_0^-$ there exists $r > 0$ such that $(\forall |\epsilon| < r)$, $L \subset V_\epsilon^-$; similarly for compact V_ϵ^+.

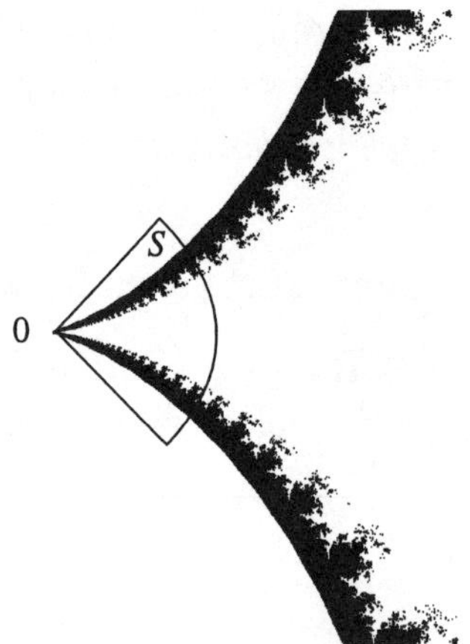

(2) V_ϵ^-/f_ϵ, V_ϵ^+/f_ϵ and $(V_\epsilon^- \cap V_\epsilon^+)/f_\epsilon$ are topological cylinders and the inclusions of $V_\epsilon^- \cap V_\epsilon^+$ in V_ϵ^- and V_ϵ^+ induce bijections of $V_\epsilon^- \cap V_\epsilon^+/f_\epsilon$ to V_ϵ^-/f_ϵ and V_ϵ^+/f_ϵ.

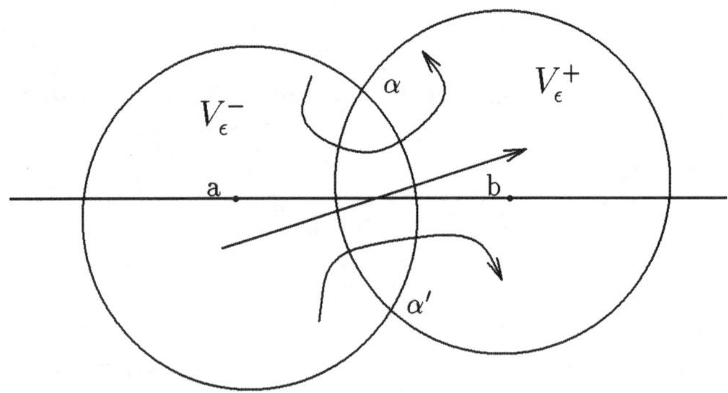

This time we do not give an explicit construction for V_ϵ^- and V_ϵ^+.

We still have Fatou coordinates $\phi_\epsilon^- : V_\epsilon^- \to \mathbf{C}$ and $\phi_\epsilon^+ : V_\epsilon^+ \to \mathbf{C}$. They are unique up to a constant, and they differ by a constant on $V_\epsilon^- \cap V_\epsilon^+$. For ϵ small enough, we have $a = \frac{-1}{8} \in V_\epsilon^-$ and $b = \frac{1}{8} \in V_\epsilon^+$; we can arrange so that holds for all $\epsilon \in S$. We normalize by $\phi_\epsilon^-(a) = \phi_\epsilon^+(b) = 0$, and we define the lifted phase $\tilde{\tau}(\epsilon)$ to be the value of $\phi_\epsilon^+ - \phi_\epsilon^-$ on $V_\epsilon^- \cap V_\epsilon^+$.

The function $\tilde{\tau}$ is holomorphic and we still have $\tilde{\tau}(\epsilon) = \frac{-\pi}{\sqrt{\epsilon}} + O(1)$.

We define the *complex phase* $\tau(\epsilon)$ as the class of $\tilde{\tau}(\epsilon)$ in $\mathbf{C}/\mathbf{Z}$.

In this setting, Prop. 19.2 holds without change. The results of §20 have been stated for $\sigma \in \mathbf{C}$, so they hold. The results of §21 are specific of the case $\sigma \in \mathbf{R}$, (ϵ_ν) a sequence in $\mathbf{R}_+$.

Let us take a sequence (ϵ_ν) in S^* which tends to 0 with $\tau(\epsilon_\nu) \to \theta \in \mathbf{C}/\mathbf{Z}$.

- If $K(f_0, \theta) = J(f_0, \theta)$, we have $K(f_{\epsilon_\nu}) \to K(f_0, \theta)$ and $J(f_{\epsilon_\nu}) \to J(f_0, \theta) = K(f_0, \theta)$.

- If $\overset{\circ}{K}(f_0, \theta)$ consists only of an attracting or virtually attracting basin, one can adapt the proof of Cor. 6.2 and prove that $K(f_{\epsilon_\nu}) \to K(f_0, \theta)$ and $J(f_{\epsilon_\nu}) \to J(f_0, \theta)$.

- In the general case, we can always extract from (ϵ_ν) a sequence such that $K(f_{\epsilon_\nu})$ and $J(f_{\epsilon_\nu})$ have limits K^*, J^*. We then have

$$J(f_0) \underset{\neq}{\subset} J(f_0, \theta) \subset J^* \subset K^* \subset K(f_0, \theta) \underset{\neq}{\subset} K(f_0).$$

This ensures that there is a discontinuity for both $f \mapsto K(f)$ and $f \mapsto J(f)$ along the sequence ϵ_ν.

23. The limit elephant

In the σ-plane, we can define $^\sigma M$ as the set of values of σ such that the critical point $\omega = \frac{-1}{2}$ of f_0 does not escape by (f_0, g_σ), i.e.

$$^\sigma M = \{\sigma \mid \omega \in K(f_0, g_\sigma)\}.$$

This set is invariant under the translation $T_1 \colon \sigma \mapsto \sigma + 1$.

On the other hand, consider the Mandelbrot set $^c M = M$, which is the connectedness locus for the family $(z^2 + c)_{c \in \mathbf{C}}$. Let W_0 be the main component of $^c\overset{\circ}{M}$ (that which contains 0). Let $^\lambda M$ be the inverse image of $^c M$ by $\lambda \mapsto c = \frac{\lambda}{2} - \frac{\lambda^2}{4}$. It is the connectedness locus of the family $(\lambda z(1 - z))_{\lambda \in \mathbf{C}}$. Corresponding to W_0 are the two discs $D_{0,1}$ and $D_{2,1}$. Making the change of variable $\mu = \frac{-2i\pi}{\lambda - 1}$, we obtain a set $^\mu M$ which is no longer compact: it contains the two half planes $\mathrm{Im}(z) > \pi$ and $\mathrm{Im}(z) < -\pi$, corresponding to the two discs. It has limbs attached at the points $x_n \pm i\pi, x_n = \frac{\pi}{\tan(\pi/n)}$, which look like elephants, and smaller elephants in between.

If we take the image $^{\tilde{\tau}} M$ of $M \cap S^*$ by $\tilde{\tau} \colon S^* \to \mathbf{C}$, we obtain essentially the same picture, because $c = \frac{1}{4} + \epsilon$ gives $\mu = \frac{-\pi}{\sqrt{\epsilon}}$ and $\tilde{\tau}(\epsilon)$ can be written as $\mu + \kappa = \eta(\mu)$, where κ is a constant, η is holomorphic, $\nu(\mu) \to 0$ and $\eta'(\mu) \to 0$ as $\mu \to \infty$; i.e. $\epsilon \mapsto 0$ in S^*.

Computer pictures show that $^\mu M$ looks more and more like $^\sigma M$ when we look more and more to the left. In order to formulate a precise statement, we need to define a topology on the set of closed (not necessarily compact) sets in $\mathbf{C}$. For $R > 0$, set $[X]_R = (X \cap D_R) \cup \partial D_R$. If (X_n) is a sequence of closed sets and $X_0 \subset \mathbf{C}$ is a closed set, we say that $X_n \to X_0$ if $(\forall R)[X_n]_R \to [X_0]_R$ as $n \to \infty$ in terms of the Hausdorff distance.

Conjecture 23.1. *The translated set $T_{\kappa + n}{}^\mu M$ tends to $^\sigma M$ as $n \to \infty$.*

Half of this conjecture can be proved rather easily:

Theorem 23.2 (Lavaurs). *For any $R > 0$,*

$$\partial([T_{\kappa + n},{}^\mu M]_R, \ [^\sigma M]_R) \to 0 \ as \ n \to \infty.$$

This is a kind of upper semi-continuity. The key point is again that escaping is an open condition.

The other half, $\partial([^\sigma M]_R, [T_{\kappa + n}{}^\mu M]_R)$ seems for the moment out of reach. One can prove that

$$(\forall R) \ \partial([\partial^\sigma M]_R, \ [T_{\kappa + n}\partial^\mu M]_R) \to 0$$

(in this formula, ∂ has two different meanings, try not to get confused!) But there is a possibility of very queer components in the interior of $^\sigma M$ which would not be limit of queer (non hyperbolic) components of $^\mu \overset{\circ}{M}$.

24. Other polynomials with parabolic cycles

Let us first consider quadratic polynomials, which can be put in the form $p_c\colon z \mapsto z^2 + c$ by an affine conjugacy. Let P_{c_0} be a polynomial with a parabolic cycle of order k, multiplier $\rho = e^{2\pi i p/q}$ (with p and q relatively prime). In the parameter plane, c_0 is the "root" of a component W of $\overset{\circ}{M}$, such that P_c has an attracting cycle of order kq for $c \in W$. Two cases must be distinguished.

Primitive case: $p = 1$, $(p/q = 0/1)$, the set W is a "primitive" component, ∂W looks like a cardioid with a cusp at c_0. The closure of each connected component of $\overset{\circ}{K}(P_{c_0})$ is a "cauliflower". When c ranges in a small sector with vertex c_0, centered on the cusp, each cauliflower undergoes modifications which are similar to those of the standard cauliflower $K(P_{1/4})$, except that there are filaments which follow roughly the real segment and its iterated preimages. These filaments keep the set connected if c is on the filment in M which leads to c_0.

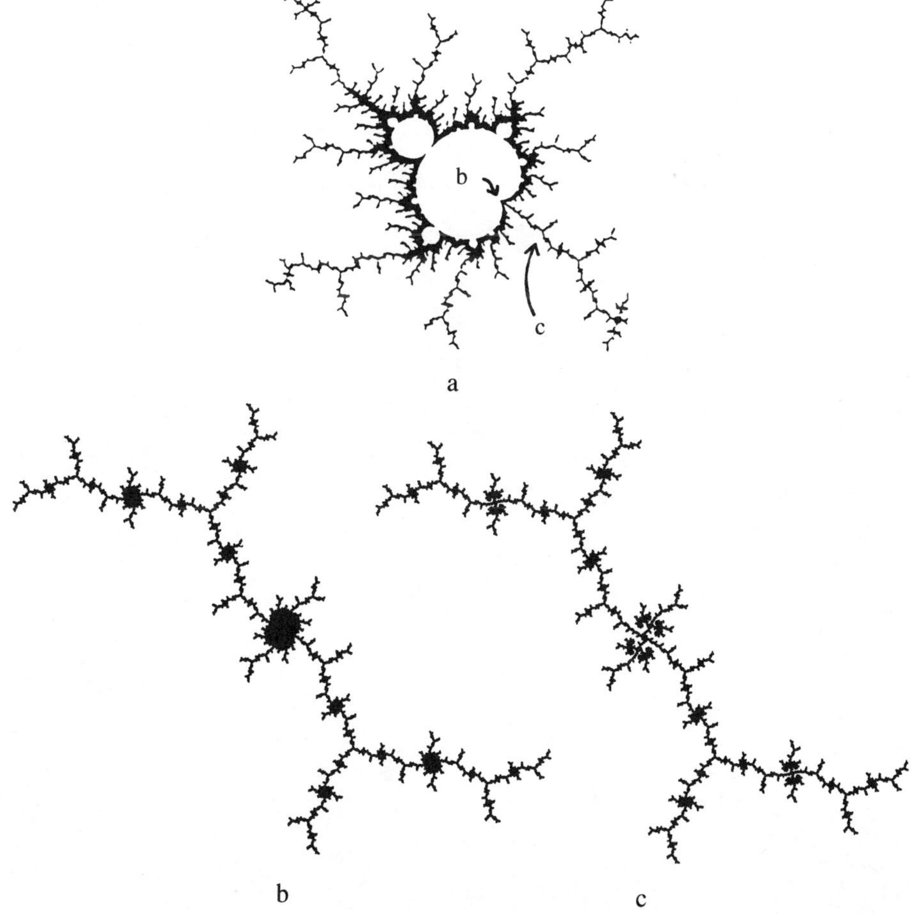

Satellite case: $\rho = e^{2\pi i p/q}$, $0 < p < q$, p and q relatively prime. The component W is attached at the point c_0 to another component W_0 where P_c has an attracting cycle of order k, and c_0 has internal argument p/q in W_0. One can draw two sectors S' and S'' with vertex c_0, centered on the half tangents common to ∂W and ∂W_0.

When c ranges in one of these sectors, we have again for each component of $\overset{\circ}{K}(P_{c_0})$ phenomena which are analogous to what occur with the standard cauliflower.

For polynomials of degree 2, the number Q of repelling axes at a parabolic point of multiplier $\rho = e^{2\pi i p/q}$ is always q. For a polynomial of higher degree, we can only assert that Q is a multiple of q. If $Q = q$, the phenomena which occurs when we move the coefficients can be analyzed by the same methods. If $Q > q$, the situation is more complicated, and has not been fully analyzed yet. But there are subfamilies which can be analyzed in the same way.

In all cases we get discontinuity at f_0 for both $f \mapsto K(f)$ and $f \mapsto J(f)$

References

[B1] P. Blanchard, "Complex analytic dynamics on the Riemann sphere," *Bull. Amer. Math. Soc.*, **11**, pp. 85-141.

[Bra] B. Branner, "The Mandelbrot set," In: *Chaos and Fractals: The Mathematics Behind the Computer Graphics*, R.L. Devaney and L. Keen (editors), Proceedings of Symposia in Applied Mathematics, Vol. 39, Amer. Math. Soc., 1989, pp. 75-105.

[Bro] H. Brolin, "Invariant sets under iteration of rational functions," *Arkiv für Mat.*, **6**, 1965, pp. 103-144.

[Cr] H. Cremer, Über der Häufigkeit der Nichtzentren," *Math. Ann.*, **115**, 1938, pp. 573-580.

[D1] A. Douady, "Systèmes dynamiques holomorphes," *Séminaire Bourbaki, Volume 1982-93, exposé no. 599, Astérisque*, **105-106**, 1983, pp. 39-63.

[D2] A. Douady, "Julia sets and the Mandelbrot set, in: [PeR].

[D3] A. Douady, "Disques de Siegel et anneaux de Hermann," *Sém. Bourbaki, Volume 1986-87, exposé no. 677, Astérisque*, **152-153**, 1987, pp. 151-172.

[D4] A. Douady, Prolongement de Mouvement holomorphes, in *Sém. Bourbaki* (Paris), Nov. 1993.

[DH1] A. Douady and J. Hubbard, "Itération des polynômes quadratiques complexes," *C.R. Acad. Sci. Paris*, **294**, 1982, pp. 123-126. [DH2] A. Douady and J.H. Hubbard, "Étude dynamique des polynômes complexes, (Première Partie)," *Publ. Math. d'Orsay 84-02*, 1984; "(Deuxième Partie)," *85-02*, 1985.

[DH3] A. Douady and J. Hubbard, "On the dynamics of polynomial-like mappings," *Ann. Sci. École Norm. Sup.*, **18**, 1985, pp. 287-343.

[Éc] J. Écalle, "Théorie itérativ: introduction a la théorie des invariants holomorphes," *J. Math. Pure Appl.*, **54**, 1975, pp. 183-258.

[Fa1] P. Fatou, "Sur les solutions uniformes de certaines équations fonctionnelles," *C.R. Acad. Sci. Paris*, **143**, 1906, pp. 546-548.

[Fa2] P. Fatou, "Sue les équations fonctionnelles," *Bull. Soc. Math. France*, **47**, 1919, pp. 161-271; **48**, 1920, pp. 33-94, pp. 208-314.

[Ju] G. Julia. "Mémoire sur l'itération des fonctions rationnelles," *J. Math. Pures Appl. (7th series)*, **4**, 1918, pp. 47-245.

[Lav] P. Lavaurs, "Systèms dynamiques holomorphes, Explosion de points périodiques parabo-
liques." Thése de Doctorat, Université Paris-Sud in Orsay (France), 1989.

[Le] L. Leau, "Étude sur les équations fonctionnelles à une ou plusièrs variables," *Ann. Fac. Sci. Toulouse*, **11**, 1897, pp. E.1-E.110.

[Lyu] M. Yu. Lyubich, "The dynamics of rational transforms: the topological picture," *Russian Math. Surveys*, **41**, 1986, pp. 43-117.

[Mi] J. Milnor, *Dynamics in One Complex Variable: Introductory Lectures*, Institute for Math. Sci., SUNY Stony Brook, 1990.

[PeR] H.-O. Peitgen and P. Richter, *The Beauty of Fractals*, Springer-Verlag, 1986.

[Sh] M. Shishikura, "The Hausdorff dimension of the boundary of the Mandelbrot set and Julia sets," *Annals of Math.*, 1993.

[Si] C.L. Siegel, "Iteration of analytic functions," *Annals of Math.*, **43**, 1942, pp. 607-612.

[St] N. Steinmetz, *Rational Iteration (Complex Analytic Dynamical Systems)*, de Gruyter (Berlin), 1993.

[Su] D. Sullivan, "Quasiconformal homeomorphisms and dynamics, I, Solution of the Fatou-Julia problem on wandering domains," *Annals of Math.*, 1985, **122**, pp. 401-418.

[Ta] Tan Lei, "Similarity between the Mandelbrot set and Julia sets," *Comm. Math. Phys.*, 1990, **134**, pp. 587-617.

[Vo] S.M. Voronin, "Analytic classification of germs of conformal mappings $(\mathbf{C}, 0) \to (\mathbf{C}, 0)$ with identity linear part," *Functional Analysis Appl.*, **15**, 1981, pp. 1-13.

[Yo] J.-C. Yoccoz, "Linéarisation des germes de difféomorphismes holomorphes de $(\mathbf{C}, 0)$," *C.R. Acad. Sci. Paris*, **306**, 1988, pp. 55-58.

COLOR PLATE I. K_c for $c = .251498$.

COLOR PLATE II. This is a detail. For $c = .251516$ one would get a similar picture, but with the butterfly-like decorations in positions alternating with the present one. For $c = .251534$ one would get a picture indistinguishable from the present one.

COLOR PLATE III. This is the corresponding detail in the limit K, i.e., in $K(c_0, g)$ for some g.

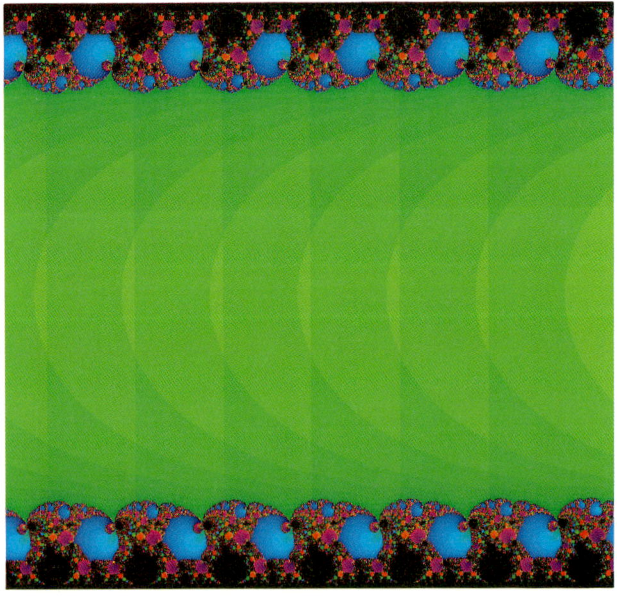

COLOR PLATE IV. This shows the limit elephants.

Proceedings of Symposia in Applied Mathematics
Volume **49**, 1994

The Dynamics of Newton's Method

Paul Blanchard

We consider the dynamics of a special class of rational functions — those rational functions that are obtained from Newton's method as applied to a polynomial equation. Such maps are interesting for two reasons:
(1) They form a natural family of non-polynomial examples; and
(2) their dynamical properties are related to their utility as numerical algorithms.

After reviewing a few basic facts, we describe a one-parameter family of degree-three rational functions derived from Newton's method applied to cubic equations in one variable. Then, in order to explain the results of a related computer experiment, we present Douady and Hubbard's remarkable theory of polynomial-like mappings.

Using Newton's method to find the roots of a polynomial equation

$$p(z) = a_d z^d + a_{d-1} z^{d-1} + \ldots + a_0 = 0$$

is identical to computing individual orbits of the dynamical system generated by the Newton map $N(z)$ (or $N_p(z)$ if we need to be explicit about the polynomial $p(z)$)

$$N(z) = z - \frac{p(z)}{p'(z)}. \tag{1}$$

As we discuss Newton's method, it is useful to keep the following somewhat typical example in mind.

1980 *Mathematics Subject Classification* (1985 *Revision*). Primary 58Fxx; Secondary 30D05.

Example 1. If we apply Newton's method to the polynomial equation $p(z) = z^3 - 1 = 0$, we obtain the rational map

$$\cdot\; N(z) = z - \frac{z^3 - 1}{3z^2} = \frac{2z^3 + 1}{3z^2}.$$

The three solutions to $z^3 = 1$ are 1 and $(-1 \pm i\sqrt{3})/2$. One can easily verify that this rational map has superattracting fixed points of multiplicity two at each of the roots. Since $\deg(N) = 3$, each root has a "third" inverse image somewhere in $\mathbb{C}$. In this example, these "prefixed" points correspond to distinct components of the Fatou set. They map injectively onto the immediate basins of the superattracting fixed points, i.e., the components of the Fatou set that contain the roots.

Note that, unlike a polynomial function, the inverse image of ∞ in this example contains more than one point. That is, $N^{-1}(\infty) = \{0, \infty\}$. In fact, as Figure 1 suggests, the preorbit of ∞ (all points that eventually map to ∞ under iteration) is dense in the Julia set. $\diamond$

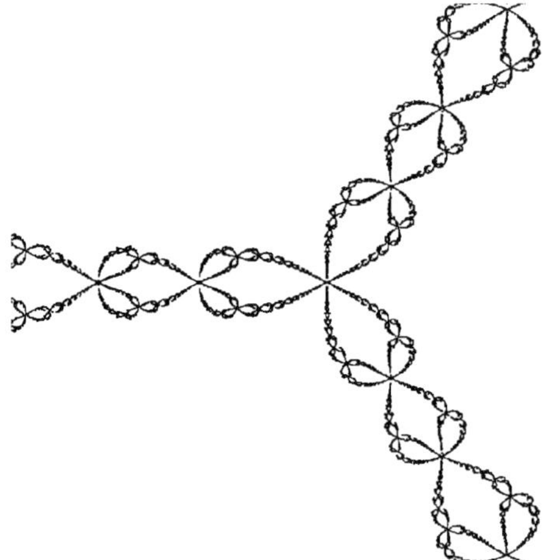

Figure 1. The Julia set of Newton's method applied to the equation $z^3 = 1$.

We begin our general discussion by summarizing a few basic facts about Newton's method for polynomials.

Remark 1.
(a) From Equation 1, we see that the roots of $p(z)$ correspond to the finite fixed points of $N(z)$.
(b) The point at infinity is a fixed point, and since $N'(\infty) = d/(d - 1)$, it is repelling. Therefore, if Newton's method produces an iterate near ∞, successive iterates tend away from infinity.
(c) The derivative of N is

$$N'(z) = \frac{p(z)p''(z)}{[p'(z)]^2},$$

and therefore, the simple roots of $p(z)$ are superattracting fixed points of $N(z)$. This is a desirable property for a root-finding algorithm because, in a neighborhood of its superattracting fixed points, the algorithm is locally conjugate

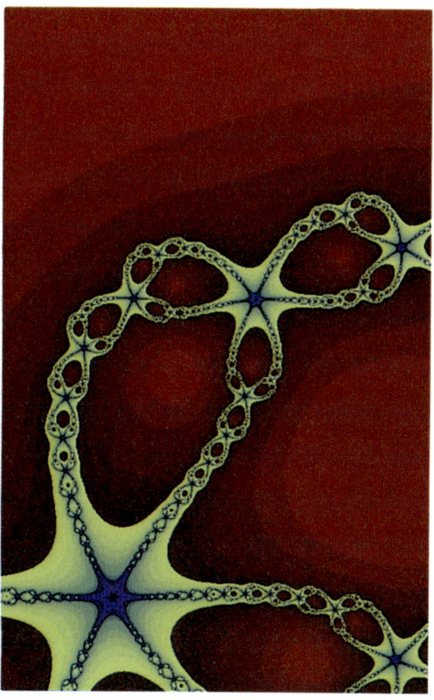

COLOR PLATE 1. An enlargement of part of the basin portrait for Newton's method applied to the equation $z^3 = 1$. The coloring scheme highlights the Julia set.

COLOR PLATE 2. The basin portrait for Newton's method applied to the equation $z^4 = 1$.

COLOR PLATE 3. The basin portrait for Newton's method applied to the equation $z^8 + 3z^4 = 4$.

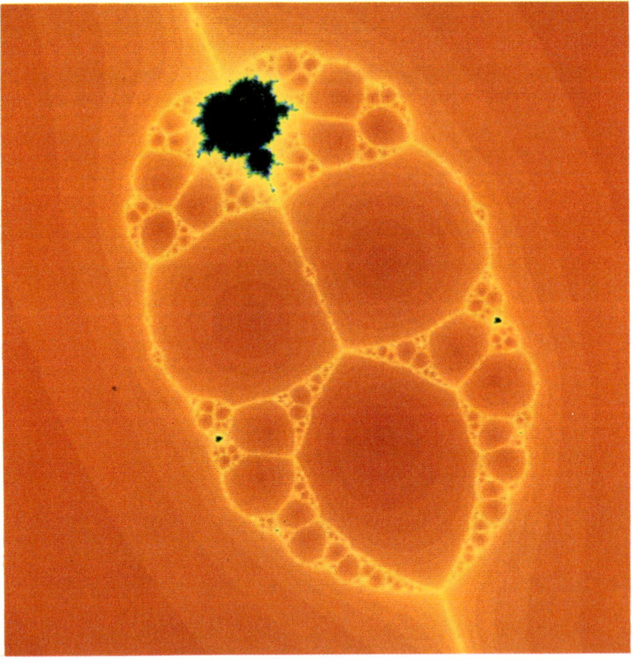

COLOR PLATE 4. An enlargement of part of the parameter space for Newton's method applied to cubic polynomials. See Figure 6 on p. 146.

to $z \mapsto z^k$ for some $k > 1$. Thus, local convergence is very rapid. In fact, the number of decimal places of accuracy at least doubles with each iteration.

(d) Multiple roots are attracting fixed points, but they are not superattracting. In fact, for a multiple root of multiplicity m, the derivative of Newton's method at the root is $(m-1)/m$. Thus, the rate of attraction is linear, and the algorithm is not very effective in this case.

(e) For a generic polynomial of degree d, the Newton's map is a rational map of degree d. However, when the polynomial has multiple roots, $\deg(N) < d$. In this section, we mostly consider polynomials that do not have multiple roots.

(f) In other chapters of this volume, we have seen that dynamical properties of a complex-analytic function are often determined by the dynamics of its critical points. Therefore, it is important to note that the critical points of $N(z)$ are the simple roots as well as the inflection points of $p(z)$.

(g) Likewise, since ∞ is a (slowly) repelling fixed point for Newton's method, we note that the poles of $N(z)$ are the critical points of $p(z)$. Consequently, orbits that avoid the critical points of $p(z)$ have the best chance of converging rapidly to a root.

(h) Unfortunately, given (f) and (g), we must keep two sets of critical points in mind. Note that it is the critical points of $N(z)$ (a subset of the roots and inflection points of $p(z)$) that predict the overall dynamics of $N(z)$.

Given Remark 1(g), one wonders how the critical points of $p(z)$ are related to its roots.

Theorem 1. (Lucas, 1874) *The critical points of $p(z)$ are contained in the convex hull of the roots of $p(z)$.* $\blacksquare$

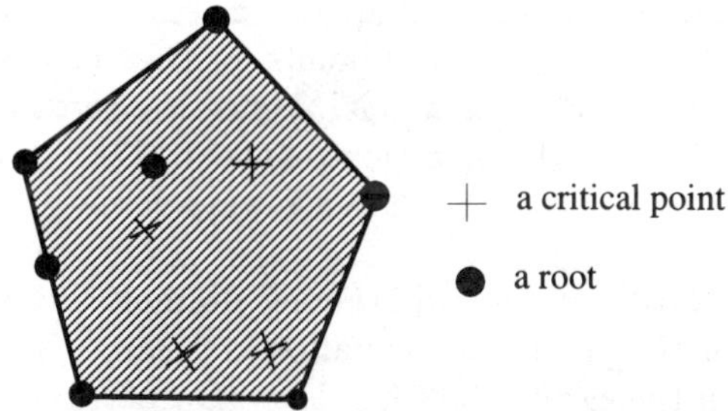

Figure 2. The critical points of $p(z)$ (i.e., the poles of $N(z)$) are contained in the convex hull of the roots of $p(z)$.

It is also useful to consider the manner in which Newton's method for a given polynomial is related to Newton's method for a "rescaled" polynomial.

Remark 2. Let $T(z) = \alpha z + \beta$ where $\alpha \neq 0$ and let $q(z) = p(T(z))$. Then

$$T \circ N_q \circ T^{-1} = N_p.$$

In other words, we can transform the roots by an affine map without qualitatively changing the dynamics of the corresponding Newton's function.

Given these general observations, we are now ready to discuss global convergence properties. Due to its simplicity, we start with the quadratic case. The result we describe has been known since at least 1870 and is explicitly discussed in the papers of Schoeder [Sch] and Cayley [C1–4] (see [A] for an interesting discussion of the history of this result).

Theorem 2. *Let $p(z)$ be a quadratic with distinct roots. Then Newton's method $N_p(z)$ is globally, analytically conjugate to the quadratic polynomial $z \mapsto z^2$.*

Proof. We can establish this result without doing any calculation. Denote the roots of the quadratic by α and β and consider the Möbius transformation

$$h(z) = \frac{z - \beta}{z - \alpha}.$$

Note that $h(\infty) = 1$, $h(\beta) = 0$, and $h(\alpha) = \infty$. Then, $h \circ N_p \circ h^{-1}$ is a rational map of degree 2 that has superattracting fixed points at 0 and ∞, and it fixes 1. It is therefore the map $z \mapsto z^2$. ∎

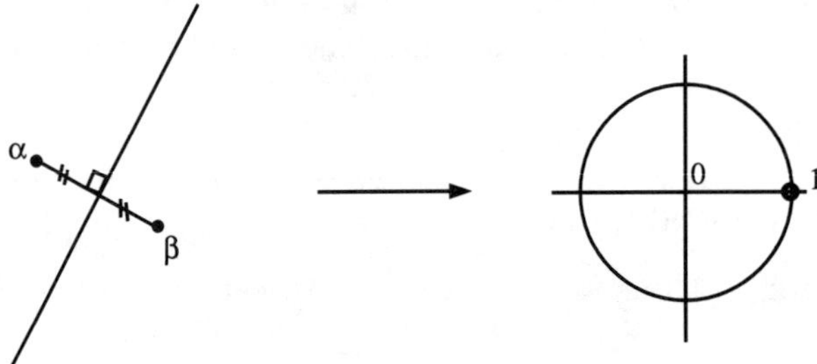

Figure 3. The conjugacy h in Theorem 2.

Note that, under the conjugacy h, the Julia set for $z \mapsto z^2$ corresponds to the perpendicular bisector of the line segment from α to β. Along this bisector, N_p has the "angle doubling" dynamics of the map $z \mapsto z^2$ restricted to the unit circle.

As we mentioned above, the critical points for Newton's method are the roots and the inflection points of the polynomial. Since the roots are always fixed, the only "free" critical points are the inflection points. In the quadratic case, there are no inflection points, and Newton's method is always conjugate to the rational map $z \mapsto z^2$.

The analysis of Newton's method becomes dramatically more complicated as soon as the degree of the polynomial equation is greater than 2. To see why, we describe work done in the early 1980's by Curry, Garnett, and Sullivan [CGS] and by Douady and Hubbard [DH]. (Tan Lei [L] has recently completed a detailed mathematical description of this parameter space.)

A Computer Experiment:

To study Newton's method for cubics using computer graphics, we use the observation in Remark 2 to eliminate duplication. Given three distinct roots in $\mathbb{C}$, there is an affine map that transforms this "triangular" configuration to one that is entirely located in the half plane $\{z \mid \operatorname{Im} z \geq 0\}$ with its longest side being the interval $[0, 1]$. Therefore, we need only consider polynomials of the form

$$p_\rho(z) = z(z - 1)(z - \rho), \tag{2}$$

where $\operatorname{Im} \rho \geq 0$, $|\rho| \leq 1$, and $|\rho - 1| \leq 1$. In other words, we use a one-parameter family of polynomials whose roots are 0, 1, and ρ, where ρ is the parameter (see

Figure 4). Moreover, by identifying the triangles in Figure 4 that are congruent via a conformal affine map, we see that this parameter space is homeomorphic to the two-sphere S^2 with one puncture. (In our computer pictures we let ρ range throughout $\mathbb{C}$.)

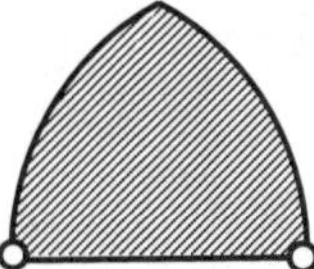

Figure 4. This figure is a sketch of the region in the ρ-plane that corresponds to a completely representative collection of cubic polynomials in the form of Equation 2. Given an arbitrary cubic q, there is a number ρ in this region such that N_q and N_{p_ρ} are globally conjugate by Remark 2.

Given this representative family of cubics, we now explore the dynamics of the associated Newton functions. In general, given a polynomial $p(z)$, the Julia set of N_p does not play an important role for two reasons. In all known examples, it has measure zero, and it usually repells nearby points. Therefore, numerical errors force most orbits to move away from the Julia set.

Consequently, we focus on the structure of the Fatou set F. Computer evidence suggests that the basins of the roots are relatively large components of F. However, F may contain basins of other attracting periodic orbits. If such orbits exist, they do not correspond to roots of $p(z)$, and every starting point in such a basin leads to an unsuccessful application of Newton's method. The following experiment ([CGS] and [DH]) is designed to locate such orbits.

Recall that all periodic attracting orbits attract at least one critical point. Thus, we use critical points to locate attracting periodic orbits. We follow the orbit of the "free" critical point of N_p, the inflection point of $p(z)$. In Figures 5 – 7, we shade the parameter value ρ according to the number of iterates it takes for the orbit of the associated inflection point to converge to one of the three roots. If it does not converge, we plot a black point at ρ.

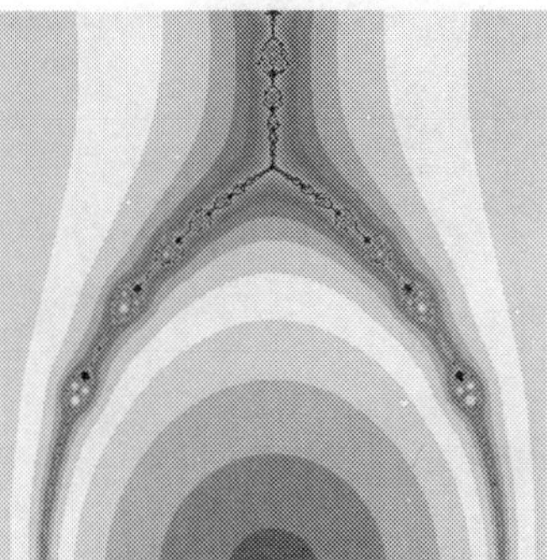

Figure 5. This figure contains the unit square

$$\{z \mid 0 \le \operatorname{Re} z \le 1,\ 0 \le \operatorname{Im} z \le 1\}$$

in the ρ-plane. Thus it contains the region sketched in Figure 4, and it is shaded using the scheme described directly above.

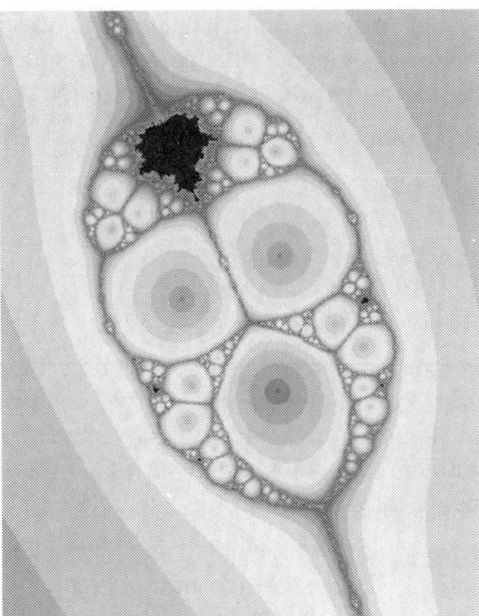

Figure 6. This figure is an enlargement of a small rectangle from Figure 5. Precise coordinates are given in the appendix. Recall that the color black corresponds to parameters for which the orbit of the inflection point does not converge to one of the three roots.

Figure 7. This figure is an enlargement of a rectangle from Figure 6. It illustrates the largest of black regions in Figure 6.

Figures 5 – 7 describe portions of parameter space. We use these figures to determine interesting values of the parameter ρ. The following two figures illustrate the structure of the Fatou set for a value of ρ chosen from the black region in Figure 7.

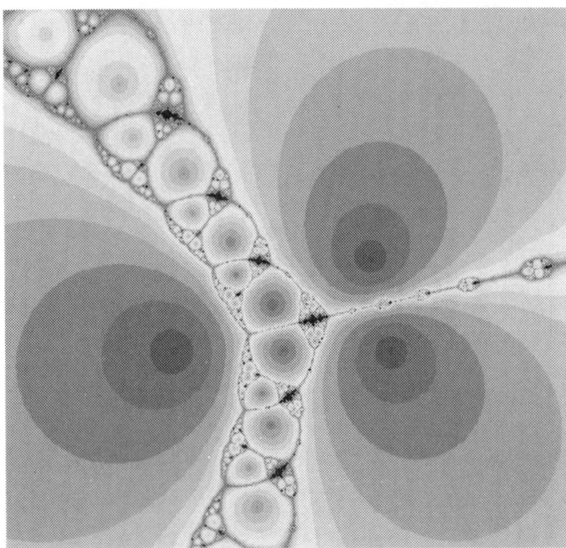

Figure 8. This picture illustrates the structure of the Fatou set for $\rho = 0.909419 + 0.416106i$. We shade a point corresponding to the number of iterates necessary for its orbit to converge to one of the roots (up to a reasonable accuracy). If the orbit does not converge after a prescribed number of iterates, a black dot is plotted at the initial point.

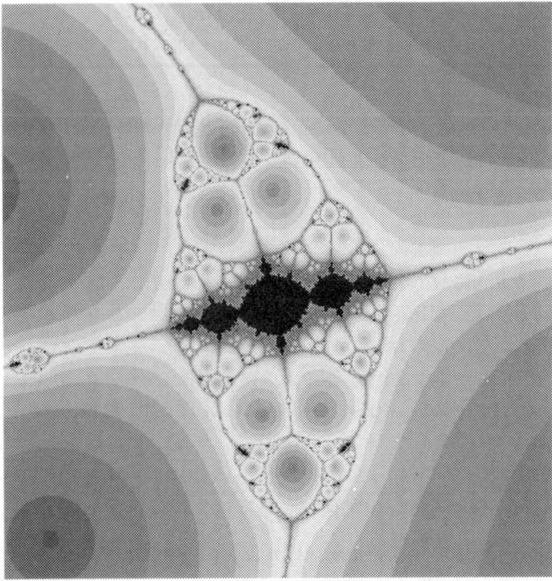

Figure 9. This figure is an enlargement of a rectangle containing one of the black regions in Figure 8. The large black region K that resembles the filled Julia set of $z \mapsto z^2 - 1$ is invariant under the second iterate of N. The union $K \cup N(K)$ contains a superattracting periodic orbit whose period is 4.

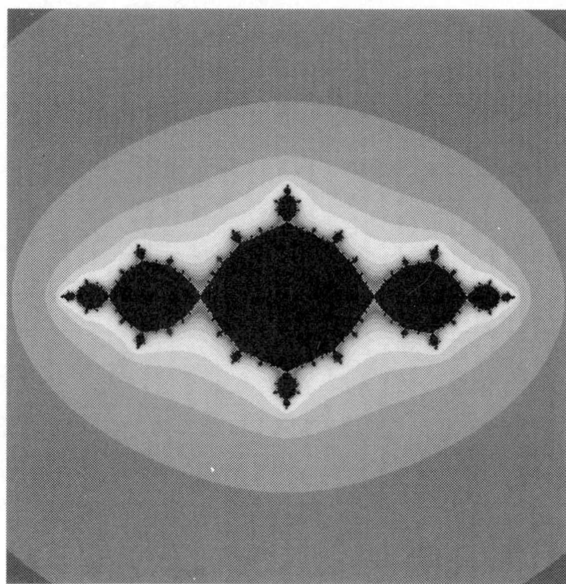

Figure 10. The filled Julia set of $z \mapsto z^2 - 1$. We include this
figure to emphasize the similarity between Figure 9 and this filled
Julia set.

The results of this experiment are remarkable. We are studying the dynamics
of maps that bear little relationship with quadratics, and we are considering a
parameter space that presumably has nothing to do with the quadratic family.
Why do we obtain images of the Mandelbrot set? What is their significance?

Douady and Hubbard [DH] answered both questions when they developed their
theory of polynomial-like mappings.

Definition. Suppose that U' and U are simply-connected domains and that U' is a
relatively compact subset of U. A map $f: U' \to U$ that is analytic and proper is
called a *polynomial-like* map.

Remarks:
(a) A map f is *proper* if the inverse image of a compact set is compact.
(b) A polynomial-like map has a finite degree which can be determined by counting
    inverse images with multiplicity.
(c) The definition of a polynomial-like map depends both on the analytic map and
    the choice of U'. For example, it is possible to find cubics that are, of course,
    cubic-like on disks of large radius around the origin but that are quadratic-like
    on smaller domains. Also, transcendental functions can be polynomial-like if
    the domain U' is chosen appropriately. See Examples 2 and 3.

Example 2. Given a cubic polynomial $p(z)$, we define a function $h_p : \overline{\mathbb{C}} \to [0, \infty]$
that is harmonic on the basin of ∞ by

$$h_p(z) = \lim_{k \to \infty} \tfrac{1}{3^k} \log_+ |p^k(z)|$$

where $\log_+ = \max\{0, \log\}$. In general, $p(z)$ has two distinct critical points c_1
and c_2. For any cubic $p(z)$ that has $c_1 \in W^s(\infty)$ and $c_2 \notin W^s(\infty)$, we obtain

a a polynomial-like map q of degree 2 by restricting $p(z)$. Let $v = h(c_1)$. Then $h(p(c_1)) = 3v$. We set $U = \overline{\mathbb{C}} - h^{-1}[3v, \infty]$ and choose U' to be the component of $p^{-1}(U)$ that contains c_2 (see Figure 11). The map $p : U' \to U$ is a polynomial-like map of degree 2. $\diamond$

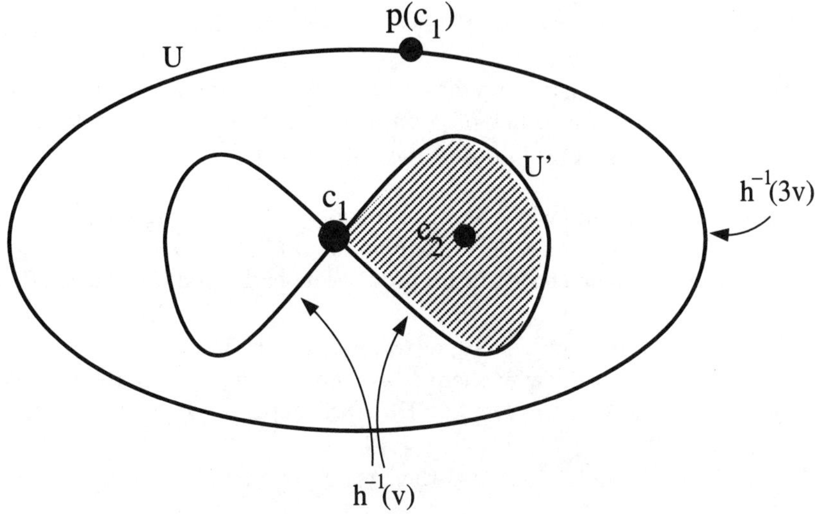

Figure 11. The levels sets of level $3v$ and v for h along with the domains U' and U.

Example 3. Let $f(z) = -2 + \cos z$ and $U' = \{z \mid |\text{Re}\, z| < 2,\ |\text{Im}\, z| < 3\}$. The map $f|U'$ is polynomial-like of degree 2 on U' even though $f(z)$ is a transcendental, entire function.

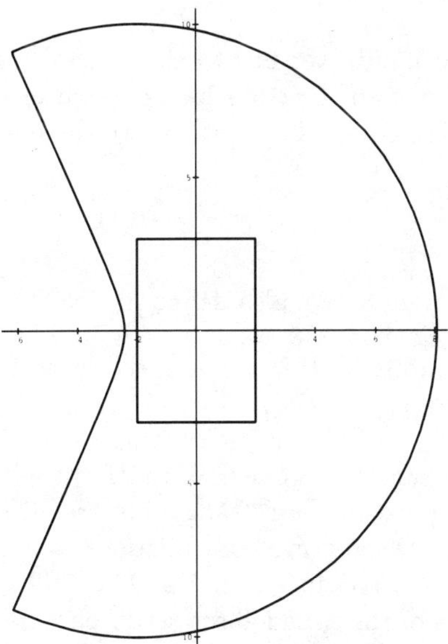

Figure 12. The rectangle U' and its image U under $f(z) = -2 + \cos z$. On U', the transcendental function $f(z)$ is polynomial-like of degree 2.

The filled Julia set of a polynomial is the set of points whose orbits do not converge to infinity. A similar concept exists for polynomial-like maps.

Definition. If $f : U' \to U$ is a polynomial-like map, then the *filled Julia set K_f* for f is

$$K_f = \{z \in U \mid f^n(z) \in U' \text{ for all } n \in \mathbb{Z}^+\}.$$

Theorem 3. (Douady-Hubbard) *Associated to each polynomial-like map f is a polynomial q such that the dynamics of f on a neighborhood of K_f is topologically conjugate to the dynamics of q on a neighborhood of K_q.* ∎

This theorem explains why the black region in Figure 9 resembles the black region in Figure 10. The map N^2 is polynomial-like of degree 2 on a region U' contained in the rectangle shown in Figure 9. The corresponding quadratic given by Theorem 3 is $z \mapsto z^2 - 1$.

Although Theorem 3 explains why we see quadratic Julia sets in Newton's method, it does not explain why there is a region in Figure 7 that resembles the Mandelbrot set. However, the Douady-Hubbard paper also contains results that apply to parameter space.

The following example illustrates their result in a familiar case.

Example 4. Let f_λ be the quadratic polynomial $f_\lambda(z) = \lambda z + z^2$. Since any quadratic is conjugate to one of the form $z \mapsto z^2 + c$, we have c as a function of λ. In fact, this function is

$$c = \frac{\lambda}{2} - \frac{\lambda^2}{4},$$

which is a branched covering map. ◇

With Example 4 in mind, we turn to the general result. Let Λ be a simply-connected domain in $\mathbb{C}$ and $\{f_\lambda \mid \lambda \in \Lambda\}$ be a one-parameter family of degree-two, polynomial-like maps. We define $\Psi \colon \Lambda \to \mathbb{C}$ where the quadratic polynomial

$$q_\lambda(z) = z^2 + \Psi(\lambda)$$

is related to f_λ by Theorem 3. We also define M_Λ as $\Psi^{-1}(M)$.

Theorem 4. (Douady-Hubbard) *If Ψ is not constant and M_Λ is compact, then the map $\Psi \colon M_\Lambda \to M$ is a branched cover.* ∎

Remark. Theorem 4 explains why we see a Mandelbrot set in Figure 7. On a simply-connected domain contained in the rectangle illustrated in Figure 7, the family of second iterates of the Newton's method functions is a polynomial-like family of degree 2 (on the appropriate domains in $\overline{\mathbb{C}}$). Thus, Theorem 4 indicates that we will see branched covers of the Mandelbrot set in parameter space.

The Douady-Hubbard paper also indicates how to determine the degree of the covering map from Theorem 4. Let A be a closed subset homeomorphic to $\overline{\mathbb{D}}$ such that $A \subset \Lambda$ and $M_\Lambda \subset \text{int}(A)$.

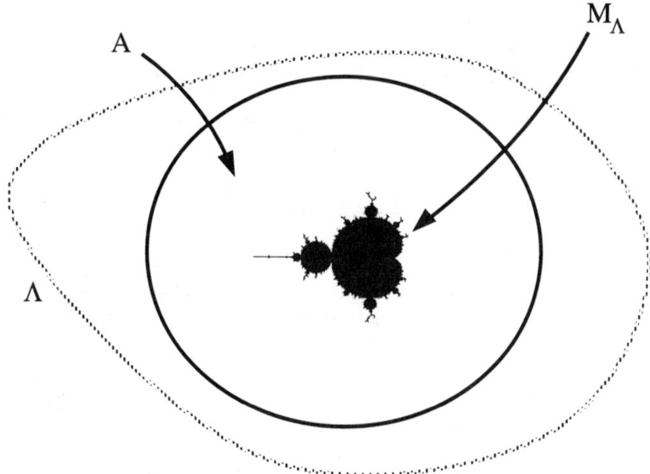

Figure 13. The closed, simply-connected region A that contains M_Λ.

Theorem 5. *Let f_λ and Ψ represent the maps in Theorem 4 and let w_λ denote the critical point of f_λ. Then, $\deg(\Psi\colon M_\Lambda \to M)$ equals the winding number of $f_\lambda(w_\lambda) - w_\lambda$ about 0 as λ traverses once around the boundary of A.* ∎

Therefore, if the winding number is 1, M_Λ is homeomorphic to the Mandelbrot set.

Polynomials whose degree is greater than three usually have move than one inflection point. Thus, examples of Newton's method with more than one periodic attracting orbit are possible. Hurley [H] has shown that, for $d \geq 3$, there is a polynomial of degree d with $d - 2$ distinct attracting periodic orbits.

Theorem 6. *(Hurley) For each $d \geq 3$, there exists a polynomial $p(z)$ of degree d whose corresponding Newton's method function has $d - 2$ distinct attracting periodic orbits of period greater than one.* ∎

Appendix

The following list provides detailed specifications for Figures 5 – 9.

Figure 5:

Lower left corner:	-0.1
Upper right corner:	$1.1 + 1.2i$

Figure 6:

Lower left corner:	$0.869039 + 0.323172i$
Upper right corner:	$0.971966 + 0.450939i$

Figure 7:

Lower left corner:	$0.894704 + 0.407193i$
Upper right corner:	$0.919835 + 0.435468i$

Figure 8:

Lower left corner:	$-0.790167 - 0.907198i$
Upper right corner:	$1.83213 \;\; + 1.55605i$

Figure 9:

Lower left corner: $0.471402 - 0.023518i$
Upper right corner: $0.807454 + 0.318906i$

References

[A] D. S. Alexander, *A History of Complex Dynamics: From Schröder to Fatou and Julia*, Vieweg, 1994.

[Ba] B. Barna, *Über die Divergenzpunkte des Newtonschen Verfahrens zur Bestimmung von Wurzeln Algebraischer Gleichungen*, Publ. Math. Debrecen, **4** (1956), 384–397.

[Be] A. Beardon, *Iteration of Rational Functions*, Springer-Verlag, Graduate Texts in Mathematics #132, 1991.

[BB] D. Berkey and P. Blanchard, *Calculus*, 3e, HBJ/Saunders College Publishing, 1992. See Appendix 4.

[B1] P. Blanchard, *Complex Analytic Dynamics on the Riemann Sphere*, Bull. Amer. Math. Soc. (New Series) **11** (1984), 85–141.

[B2] P. Blanchard, *Disconnected Julia sets*. Book chapter in *Chaotic Dynamics and Fractals*, ed. by Barnsley and Demko, Academic Press, 1986, 181–201 .

[BC] P. Blanchard and A. Chiu, *Complex Dynamics: An Informal Discussion*. Book chapter in *Fractal Geometry and Analysis*, ed. by Jacques Bélair and Serge Dubuc, Kluwer Academic Publishers, 1991, 45–98.

[BL] P. M. Bleher and M. Yu. Lyubich, *The Julia Sets and Complex Singularities in Hierarchical Ising Models*, preprint #1990/12, SUNY Stony Brook, Institute for Mathematical Sciences.

[Br] H. Brolin, *Invariant sets under iteration of rational functions*, Arkiv für Matematik **6** (1965), 103–144.

[C1] A. Cayley, *The Newton-Fourier Imaginary Problem*, Amer. J. Math **II** (1879), 97.

[C2] A. Cayley, *On the Newton-Fourier Imaginary Problem*, Proc. Cambridge Phil. Soc. **3** (1880), 231–232.

[C3] A. Cayley, *Application of the Newton-Fourier Method to an Imaginary Root of an Equation*, Quart. J. of Pure and Applied Math. **XVI** (1879), 179–185.

[C4] A. Cayley, *Sur les Racines D'une Équation Algébrique*, Comptes Rendus Acad. Sci. Paris **t.CX** (Janvier–juin 1890), 174–176, 215–218.

[CGS] J. Curry, L. Garnett and D. Sullivan, *On the iteration of a rational function: Computer experiments with Newton's method*, Comm. Math. Phys. **91** (1983), 267–277.

[De] R. Devaney, *An Introduction to Chaotic Dynamical Systems*, 2 ed., Addison-Wesley, 1989.

[D1] A. Douady, *Systèmes Dynamiques Holomorphes*, Séminaire Bourbaki 1982/1983, Exposé 599, *Asterisque* **105–106** (1983), Societé math. de France.

[D2] A. Douady, *Julia sets and the Mandelbrot set, The Beauty of Fractals*, by Peitgen and Richter, Springer-Verlag, 1986, 161–173.

[D3] A. Douady, *Chirurgie sur les applications holomorphes*, Proceedings of the I.C.M., Berkeley (1986), Amer. Math. Soc., 724–738.

[DH] A. Douady and J.H. Hubbard, *On the dynamics of polynomial-like mappings*, Ann. Sci. Ec. Norm. Sup. (Paris) **18** (1985), 287–343.

[DM] P. Doyle and C. McMullen, *Solving the quintic by iteration*, Acta Math. **163** (1989), 151–180.

[EL] A. E. Eremenko and M. Yu. Lyubich, *The dynamics of analytic transformations*, Leningrad Math. J. **1** (1990), 563–634.

[F1] P. Fatou, *Sur les équations fonctionelles*, Bull. Soc. Math. France **47** (1919), 161–271; **48** (1920), 33–94 and 208–314.

[F2] P. Fatou, *Sur les solutions uniformes de certaines équations fonctionnelle*, C.R. Acad. Sci. Paris **143** (1906), 546–548.

[FS] M. Flexor and P. Sentenac, *Algorithmes de Newton généralises*, CRAS **308** (1989), 445–448.

[Fr] J. Friedman, *On the convergence of Newton's method*, Journal of Complexity **5** (1989), 12–33.

[HP] F. von Haeseler and H.-O. Peitgen, *Newton's method and complex dynamical systems*, Acta Appl. Math. **13** (1988), 3–58.

[H] M. Hurley, *Multiple Attractors in Newton's Method*, Ergodic Theory Dynamical Systems **6** (1986), 561–569.

[HM] M. Hurley and C. Martin, *Newton's algorithm and chaotic dynamical systems*, SIAM J. Math. Anal. **15** (1984), 238–252.

[J] G. Julia, *Memoire sur l'itération des fonctions rationnelles*, J. Math. pures et app. **8** (1918), 47–245. See also *Oeuvres de Gaston Julia*, Gauthier-Villars, Paris **1**, 121–319.

[K] L. Keen, *Julia Sets, Chaos and Fractals, the Mathematics behind the Computer Graphics*, ed. by Devaney and Keen, Proc. Symp. Appl. Math. **39**, Amer. Math. Soc., 1989, 57–75.

[Kr] H. Kriete, *On the Efficiency of Relaxed Newton's Method*, preprint (1992), Ruhr-Universität Bochum.

[L] T. Lei, *Cubic Newton's method of Thurston's type*, preprint, Laboratoire de Mathématiques, École Norm. Sup. de Lyon.

[Ly1] M. Lyubich, *The dynamics of rational transforms: the topological picture*, Russian Math. Surveys **41:4** (1986), 43–117.

[Ly2] M. Lyubich, *Some typical properties of the dynamics of rational maps*, Russian Math. Surveys **38** (1983), 154–155.

[Ly3] M. Lyubich, *An analysis of the stability of the dynamics of rational functions*, Selecta Math. Sovietica **9** (1990), 69–90. (Russian original published in 1984.)

[M]   B. Mandelbrot, *The Fractal Geometry of Nature*, Freeman, 1982.

[Ma] A. Manning, *How to be sure of finding a root of a complex polynomial using Newton's method*, Bol. Soc. Mat. **22** (1992), 157–177.

[Mc] C. McMullen, *Families of rational maps and iterative root-finding algorithms*, Ann. of Math. **125** (1987), 467–493.

[Mi] J. Milnor, *Dynamics in One Complex Dimension: Introductory Lectures*, preprint
#1990/5, SUNY Stony Brook, Institute for Mathematical Sciences.

[PR] H.-O. Peitgen and P.H. Richter, *The Beauty of Fractals*, Springer-Verlag, 1986.

[PSH] H.-O. Peitgen, D. Saupe, and F.v. Haeseler, *Cayley's problem and Julia sets*, Math. Intelligencer **6**, 11–20.

[Pr] F. Przytycki, *Remarks on the simple connectedness of basins of sinks for iterations of rational maps*, Dynamical Systems and Ergodic Theory, ed. by K. Krzyzewski, Polish Scientific Publishers, Warsaw, 1989, 229–235.

[SU] D. Saari and J. Urenko, *Newton's method, circle maps and chaotic motion*, Amer. Math. Monthly, **91** (1984), 3–17.

[S1] D. Sullivan, *Quasiconformal Mappings and Dynamics I*, Annals of Mathematics **122** (1985), 401–418.

[S2] D. Sullivan, *Conformal dynamical systems*, Geometric Dynamics, ed. Palis, Lecture Notes in Math. **1007**, Springer-Verlag, 1983, 725–752.

[Sch] E. Schröder, *Ueber iterirte Functionen*, Math. Ann. **3** (1871); see p. 303.

[Sh] M. Shishikura, *On the quasiconformal surgery of rational functions*, Ann. Sci. Éc. Norm. Sup. **20** (1987), 1–29.

[Su] S. Sutherland, *Finding Roots of Complex Polynomials with Newton's Method*, (preprint) Institute for Mathematical Sciences, SUNY Stony Brook.

[SVB] S. Sutherland, G. Vetger and P. Blanchard, *Citool*, computer software for Sun workstations available via anonymous ftp from math.sunysb.edu.

Department of Mathematics
Boston University
Boston, MA 02215
paul@math.bu.edu

Proceedings of Symposia in Applied Mathematics
Volume **49**, 1994

The Spider Algorithm

JOHN H. HUBBARD AND DIERK SCHLEICHER

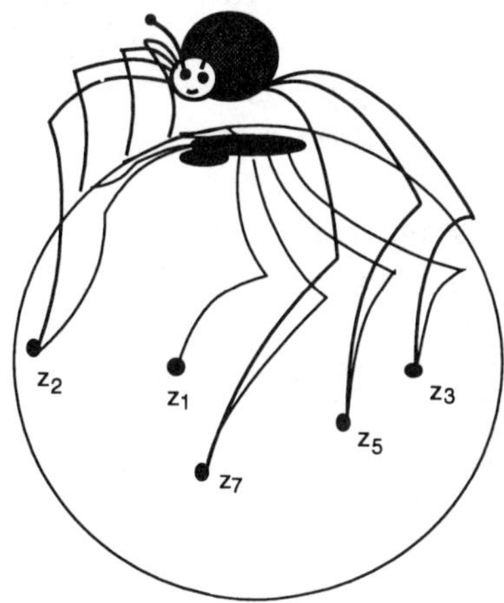

Charlotte [W] casts a 43/255-shadow.

One of the reasons complex analytic dynamics has been such a successful subject is the deep relation that has surfaced between conformal mapping, dynamics and combinatorics. The object of the spider algorithm is to construct polynomials with assigned combinatorics.

This shows up when you try to understand the Mandelbrot set. For this discussion we will write our quadratic polynomials $Q_c(z) = z^2 + c$. Every such polynomial has a *filled in Julia set* K_c, formed of the points with bounded orbits under iteration of Q_c.

A result of Fatou asserts that if the critical point $0 \in K_c$, then K_c is connected, and if $0 \notin K_c$, then K_c is a Cantor set. By definition, the Mandelbrot set M is the set of c for which K_c is connected.

Let $\mathbb{D}$ denote the open unit disc, and let $\Phi_M : \overline{\mathbb{C}} - M \to \overline{\mathbb{C}} - \mathbb{D}$ be the conformal mapping which maps ∞ to ∞ and is tangent to the identity at infinity. The existence of this mapping is not obvious; it is proved to exist at the same time as the

Supported by NSF grant DMS-9206924

AMS 1991 Mathematics Subject Classification numbers 34C35, 30D05, 34A20, 32G15

Mandelbrot set M is shown to be connected. Now call

$$R_M(\theta) = \Phi_M^{-1}\left(\{re^{2\pi i\theta} \mid r > 1\}\right)$$

the *external ray of M at angle* θ. When the limit

$$\lim_{r \searrow 1} \Phi_M^{-1}(re^{2\pi i\theta})$$

exists, we say that the ray at angle θ *lands* at the limit point.

Something which sounds like magic occurs: If θ is rational, then the ray at angle θ does land at some point c_θ, and the dynamics of Q_{c_θ} reflects the digits of θ written in base 2. This is proved in [DH1]; a simpler proof can be found in [S]. A couple of examples should bring out what this means.

Example 1. Consider the polynomial $z^2 + i$, which is at the end of the ray at angle $1/6$ turn. The number $1/6$ is written in base 2

$$1/6 = .0 \qquad 01 \quad 01 \quad 01 \quad 01 \quad \ldots$$

and under iteration of $z \mapsto z^2 + i$, the orbit of 0 is

$$i \qquad -1+i \quad -i \quad -1+i \quad -i \quad -1+i \quad -i \quad \ldots$$

Both after one term repeat with period 2.

Example 2. The external rays of M at angles $1/3$ and $2/3$ land at -0.75, the root of the component of $\overset{\circ}{M}$ in which all polynomials have an attractive cycle of period 2;

$$1/3 = .\overline{01} \quad \text{and} \quad 2/3 = .\overline{10}$$

are the numbers which in base 2 have digits which repeat with period 2.

Similarly, the rays at angle $3/7$ and $4/7$ land at $-7/4$, which is the root of a component of $\overset{\circ}{M}$ in which the polynomials have an attractive cycle of period 3. Note that

$$3/7 = .\overline{011} \quad \text{and} \quad 4/7 = .\overline{100} \quad ;$$

both have digits which repeat with period 3. This pattern extends to all hyperbolic components of the interior of M (presumably, all components are hyperbolic).

These examples illustrate the sort of combinatorics polynomials may have under iteration. We will restrict our attention to polynomials for which the critical point has a finite orbit under iteration, which means it becomes eventually periodic. We will be interested in the lengths of periodic cycles, and also in the order in which points of a cycle appear. This has an obvious meaning when the polynomials and cycles are real (see Section 1) and the appropriate generalization to the complex case can be found in Section 2.

The spider algorithm is an answer both to the question of which combinatorial patterns are realized by polynomials and how to find them. The finiteness of the critical orbit gives an algebraic equation, but this equation does not distinguish the combinatorics of its many solutions. For instance, if you look for quadratic polynomials $P_c(z) = z^2 + c$ for which the critical point is periodic of period p, where p is prime, you land on an equation of degree $2^{p-1} - 1$ after removing the trivial factor, but the algebra does not distinguish between the solutions.

The underlying techniques of the spider algorithm are due to Bill Thurston and have been elaborated on by many others: Ben Bielefeld, Adrien Douady, Yuval Fisher, Lisa Goldberg, Janet Head, Silvio Levi, Jiaqi Luo, Jack Milnor, Curt Mc-Mullen, Alfredo Poirier, Mary Rees, Mitsuhiro Shishikura, Dennis Sullivan, Tan Lei, Ben Wittner, and no doubt others to whom we owe apologies. Yuval Fisher and Ben Bielefeld have written programs implementing the spider algorithm.

Thurston's theorem reduces the problem of finding a rational function with assigned combinatorics to finding a fixed point for a mapping in an appropriate Teichmüller space. He shows that either such a fixed point exists and is unique, or there is a *Thurston Obstruction*, which consists of a set of simple closed curves having special properties. The proof, written in [DH2], is deep and difficult.

In Section 7 we will give a proof of a special case in which there is no obstruction. This proof is much easier and can serve as an introduction to the general case. We also show with an example how an obstruction can prevent such a fixed point from existing.

Let the reader be reassured: this paper does not require any knowledge of Teichmüller theory, or quasiconformal mappings, etc. The Teichmüller space is replaced by the (more intuitive) space of spiders, and the Thurston mapping by the spider map.

1. REAL KNEADING SEQUENCES AND QUADRATIC POLYNOMIALS

Just what it means for a polynomial to have assigned combinatorics is delicate to define, but for real quadratic polynomials such that the orbit of the critical point is finite this is quite easy to understand, and even to program.

Suppose that we want to find a quadratic polynomial, normalized to the form $x^2 + c$, which realizes the combinatorics sketched in the following picture.

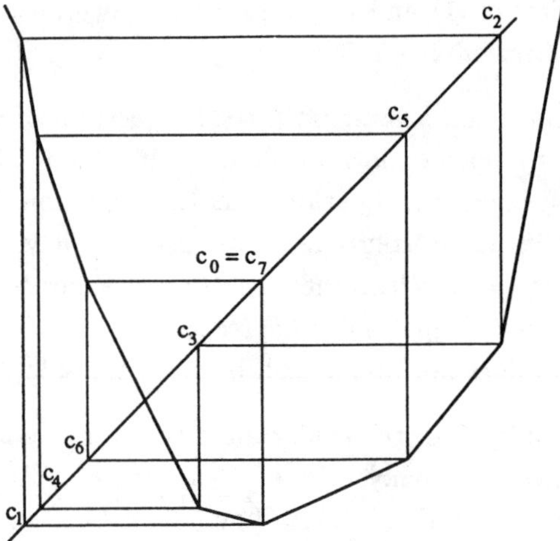

More generally, you could start with any continuous real function which has a unique minimum, which is strictly monotone on both sides of the minimum, and for which the forward orbit of the minimum is finite. The task is to find a quadratic polynomial for which the critical point will have a finite orbit with the same number of points, and such that these points will appear in the same order.

In our case, we want a real quadratic polynomial P whose critical point c_0 is periodic of period 7, and such that the points $c_j = P^j(c_0)$ appear in the order

$$c_1 < c_4 < c_6 < c_3 < c_0 = c_7 = 0 < c_5 < c_2$$

(here, P^j stands for the j-th iterate of P).

The name of the game will be to choose correct square roots; the necessary combinatorial information will be coded in what is called a *kneading sequence*, which says for every point on the critical orbit $c_1, c_2, \ldots$ if it is on the left, on the right, or equal to the critical point. Labeling these cases by L, R, and C respectively, we obtain in our case the periodic kneading sequence

$$\overline{L\,R\,L\,L\,R\,L\,C}.$$

In this setting, called "real unimodal", one can verify that there is at most one order on the points of the orbit which is compatible with a kneading sequence. This means that most orders of points on the orbit are not permissible; if one order is permissible, it carries no more information than the kneading sequence. The analogous statement will not be true for the complex case.

Consider the set of all sequences of real numbers

$$x_1 < x_4 < x_6 < x_3 < x_0 = x_7 = 0 < x_5 < x_2.$$

We will associate to such a sequence a new sequence

$$\tilde{x}_1 < \tilde{x}_4 < \tilde{x}_6 < \tilde{x}_3 < \tilde{x}_0 = \tilde{x}_7 = 0 < \tilde{x}_5 < \tilde{x}_2$$

defined as follows: let P be the polynomial $x^2 + x_1$, and let $\tilde{x}_j$ be the point of $P^{-1}(x_{j+1})$ which is to the left or the right of 0, according to whether the jth term of the kneading sequence is L or R (for $j = 1, \ldots, 6$). As the only inverse image of x_1 is 0, we obtain $\tilde{x}_0 = \tilde{x}_7 = 0$ in accordance with the entry C in the kneading sequence. This procedure is illustrated in the following picture.

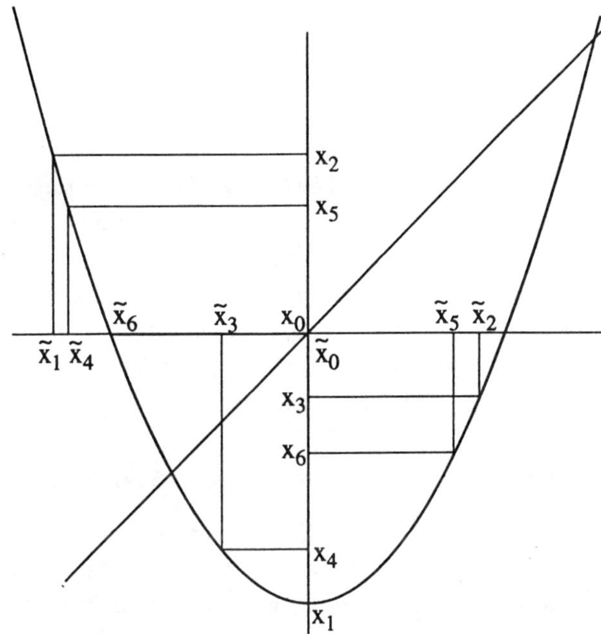

We will call the mapping

$$(x_1, \ldots, x_7) \mapsto (\tilde{x}_1, \ldots, \tilde{x}_7)$$

the real spider map associated to the combinatorial data. It should be clear that if it has a fixed point $(x_1, \ldots, x_7)$, then the polynomial $P(x) = x^2 + x_1$ is the answer to our question. In the particular case above, it does converge, and the limiting x_1 is approximately $-1.674066\ldots$. The graph of that polynomial, with the critical orbit drawn in, is represented in the following figure.

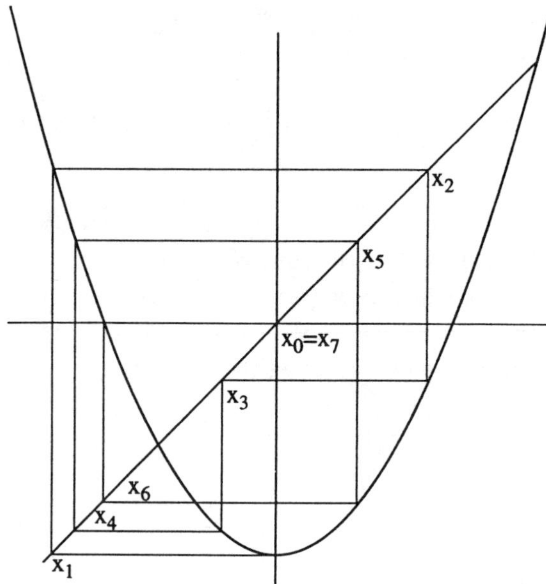

It is not too difficult to show, using the Brouwer fixed point theorem, that the real spider map has a fixed point, at least if you allow some of the points to coalesce. As far as we know, no one has been able to prove the uniqueness of the fixed point by purely real methods; complex analysis appears to be required. In the next section we will set up an analog of the spider mapping above in a complex context; a proof of its convergence in the case where the critical point is periodic, which applies in particular to the case above, appears in Section 7.

2. Complex Kneading Sequences and Standard Spiders

In this section, we will describe generalizations of the spaces of ordered sequences $(x_1, \ldots, x_n)$ and the kneading sequences above. Since sequences of complex numbers do not have a natural order, "left" and "right" do not make sense, so the notions will be somewhat more elaborate.

A new convention. We will break the usual convention of writing quadratic polynomials as $z^2 + c$, and will instead write them as

$$P_\lambda(z) = \lambda \left(1 + \frac{z}{2}\right)^2 .$$

In this normalization, the critical point of P_λ is -2, and the critical value is 0; for our purposes, this works better than the usual normalization, which is focused on the behavior at infinity. It also generalizes nicely to

$$\lambda \left(1 + \frac{z}{d}\right)^d$$

(critical point $-d$, critical value 0), and ultimately to λe^z. The spider construction generalizes to all of these, and introduces unexpected correspondences between polynomials of different degree. Investigating these correspondences is a lot of fun.

We will restrict ourselves to degree 2, although everything goes through essentially without change for the above mentioned polynomials of any degree with a single critical point, and can be adapted to go through in much greater generality; see Section 8.

Recall that we only deal with polynomials for which the critical point has a finite orbit under iteration; the corresponding condition on angles θ is that they have a finite forward orbit under angle doubling, which is the same thing as saying that the angles are rational. As we will see, there are two quite different cases which occur: the case where θ is periodic under angle doubling, and the case where it is preperiodic. For instance, 4/15 is periodic of period 4:

$$\frac{4}{15} \mapsto \frac{8}{15} \mapsto \frac{1}{15} \mapsto \frac{2}{15} \mapsto \frac{4}{15} \quad \text{(we are working in } \mathbb{R}/\mathbb{Z}\text{, i.e., mod 1);}$$

but 1/6 is not periodic: under angle doubling, we have

$$\frac{1}{6} \mapsto \frac{1}{3} \mapsto \frac{2}{3} \mapsto \frac{1}{3}\ldots \quad \text{which never returns to } \frac{1}{6}.$$

Let $\mathbb{T} = \mathbb{R}/\mathbb{Z}$ be the set of angles, counted in full turns, and let $\mathbb{T}^{\mathbb{Q}} = \mathbb{Q}/\mathbb{Z}$ be the rational ones among them.

Exercise. Show that for every $\theta \in \mathbb{T}^{\mathbb{Q}}$ there are unique smallest integers $l \geq 0$ and $k \geq 1$ and an integer a such that

$$\theta = \frac{a}{2^l(2^k - 1)}$$

(this fraction is not necessarily in lowest terms). Show that the binary expansion of θ has exactly l preperiodic digits, after which it becomes periodic of period k; similarly, the sequence θ, 2θ, $2^2\theta$, $2^3\theta$, ... becomes, after exactly l steps, periodic with period k. Periodic angles are exactly those which, when written as a fraction in lowest terms, have odd denominator.

All our angles θ will be rational. l and k will always denote preperiod and period, respectively; we will also set $n = k + l$ throughout. For periodic angles, n will be equal to the period length.

Given a rational angle $\theta \in \mathbb{T}^{\mathbb{Q}}$, we will build two combinatorial objects: The first is the standard θ-spider $\mathbb{S}_\theta \subset \overline{\mathbb{C}}$, which is the set

$$\left\{ re^{2\pi i 2^{j-1}\theta} \,|\, r \geq 1, j = 1, 2, \ldots \right\} \cup \{\infty\}.$$

We might think of this as a daddy-long-legs (as in the first picture of the paper) with its body out at infinity, and with legs streching out to the unit circle. Note that the number of legs of the spider $\mathbb{S}_\theta$ is $n = k + l$ and in particular finite. For the endpoints of the legs of the standard spider $\mathbb{S}_\theta$ we will write $x_j = e^{2\pi i 2^{j-1}\theta}$.

The other piece of combinatorial information is the *kneading sequence* of θ. Cut $\mathbb{T}$ at the halves of θ: $\theta/2$ and $(\theta+1)/2$. Label A the open component of $\mathbb{T}$ which contains θ, and the other B. Now the *θ-itinerary* $k_\theta(\alpha)$ of an angle α is the sequence of symbols $a_1, a_2, \ldots$, where

$$a_j = \begin{cases} A & \text{if } 2^{j-1}\alpha \in A \\ B & \text{if } 2^{j-1}\alpha \in B. \end{cases}$$

It may happen that one of the angles $2^{j-1}\alpha$ equals one of the boundary points $(\theta+1)/2$ and $\theta/2$. In these cases, we use the labels $*_1$ for the point which is at the counterclockwise end of A, and $*_2$ for the other.

The kneading sequence of θ is $K(\theta) = k_\theta(\theta)$. It contains one of the symbols $*_i$ if and only if θ is periodic under angle doubling.

Example 3. Consider $\theta = 9/56$. Then the standard θ-spider $\mathbb{S}_{9/56}$ is the following graph.

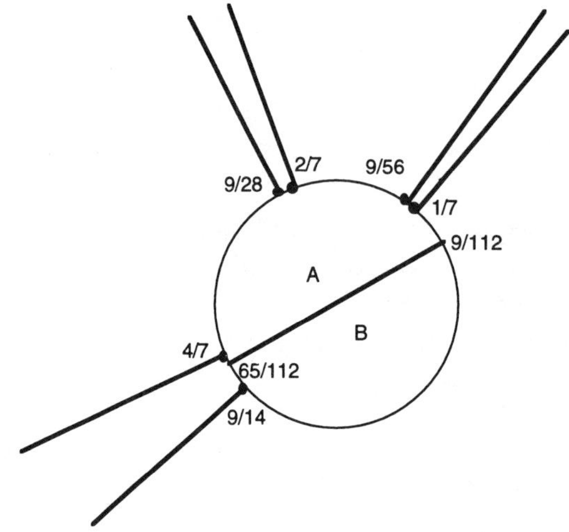

The diameter connects the two halves of 9/56, which are 9/112 and 65/112, so we can read off the kneading sequence:

$$K(9/56) = AAB\,\overline{AAA}.$$

Example 4. Let $\theta = 4/15$, which is periodic of period 4. The kneading sequence is then $K(4/15) = \overline{AAB*_2}$.

Exercise. Check that, if the angle θ is periodic under angle doubling, the kneading sequence $K(\theta)$ is periodic, too. If the angle θ is strictly preperiodic, then the kneading sequence $K(\theta)$ will also be strictly preperiodic, such that the number of steps before the period is equal. The length of the periodic part of the kneading sequence divides that of the angle.

Remark. Example 3 above shows that, for preperiodic angles, the periodic part of the kneading sequence may in fact be strictly shorter than the periodic part of the binary expansion of the angle. Irrational angles (whose binary expansions are not eventually periodic) may or may not have periodic kneading sequences.

3. Spiders and the Spider Map

Define the space $\mathcal{S}_\theta$ of θ-spiders to be the quotient of the space

$$\mathcal{S}_\theta^0 = \left\{ \begin{array}{c} \varphi : \mathbb{S}_\theta \to \overline{\mathbb{C}} \mid \varphi(\infty) = \infty, \varphi(x_1) = 0, \varphi \text{ injective, continuous} \\ \text{and respects the circular order at } \infty. \end{array} \right\}$$

by the equivalence relation where two θ-spiders φ_0, φ_1 are equivalent if there exists a continuous 1-parameter family φ_t of θ-spiders connecting them such that for every $j \geq 2$ the ratio $\varphi_t(x_j)/\varphi_t(x_2)$ is constant as a function of t.

Our equivalence relation is really generated by two different equivalence relations: moving the legs with endpoints fixed, and scaling Note that the standard spider $\mathbb{S}_\theta$ is not in the spider space, because the endpoint $\varphi(x_1)$ is not at the origin.

In the following figure, all three spiders are elements of $\mathcal{S}_{9/56}^0$; while A and B are equivalent, C is different (the legs to z_3 and z_5 cannot be untangled without moving the endpoints).

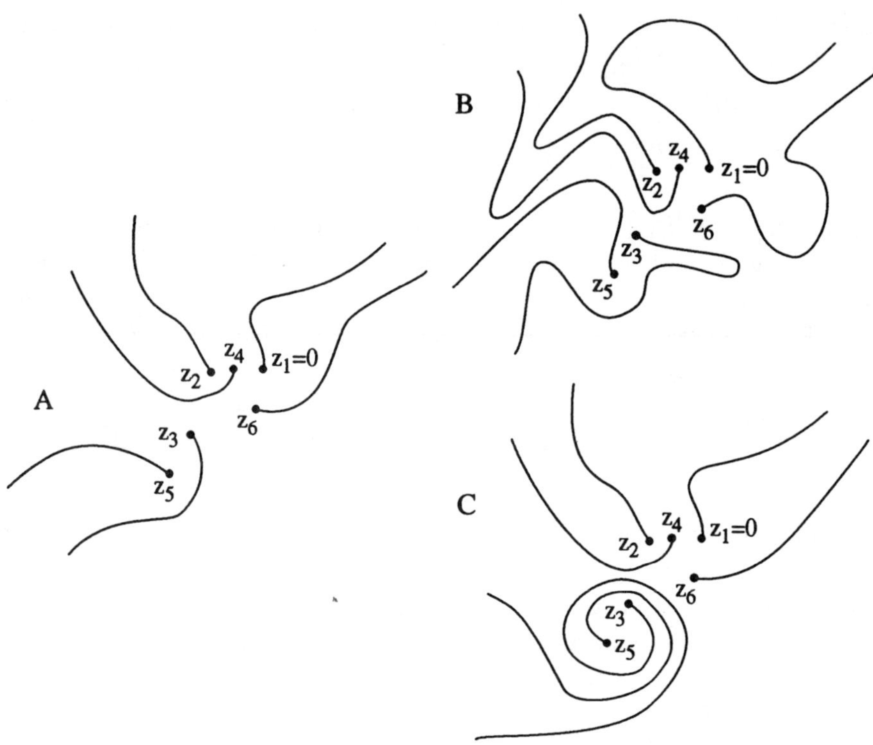

Now the principal actor of this paper can be introduced: the spider mapping

$$\sigma_\theta : \mathcal{S}_\theta \to \mathcal{S}_\theta.$$

Let $\varphi \in \mathcal{S}_\theta^0$, and call $z_j = \varphi(x_j)$ and γ_j the leg leading to z_j; we will add a tilde to the corresponding parts of $\tilde{\varphi} = \sigma_\theta(\varphi)$. Let P be the polynomial

$$P(z) = z_2 \left(1 + \frac{z}{2}\right)^2.$$

The set $P^{-1}(\gamma_1)$ is a curve cutting the plane into two parts, because γ_1 is a curve going from ∞ to 0 (the critical value of P), so the two lifts of γ meet at the critical point -2. Label A and B the two sectors of the plane obtained, so that 0 is in A.

Now define $\tilde{z}_j$ to be the point in $P^{-1}(z_{j+1})$ which is in the sector specified by the jth term of the kneading sequence $K(\theta)$, and let $\tilde{\gamma}_j$ be the component of $P^{-1}(\gamma_{j+1})$ which has $\tilde{z}_j$ as an endpoint.

Exercise. Check that we obtain exactly n disjoint legs again, which have the same circular order near ∞ as before.

Exercise. Verify that the equivalence class of $\tilde{\varphi}$ depends only on the equivalence class of φ, so that the spider map is well defined as a map from $\mathcal{S}_\theta$ to itself.

The following picture illustrates this procedure for $\theta = 9/56$. On the right is the spider φ, on the left the spider $\tilde{\varphi}$, where we have chosen the inverse image in the sector A or B according to the kneading data. Note that both $\tilde{z}_3$ and $\tilde{z}_6$ are inverse images of z_4, since $z_7 = z_4$.

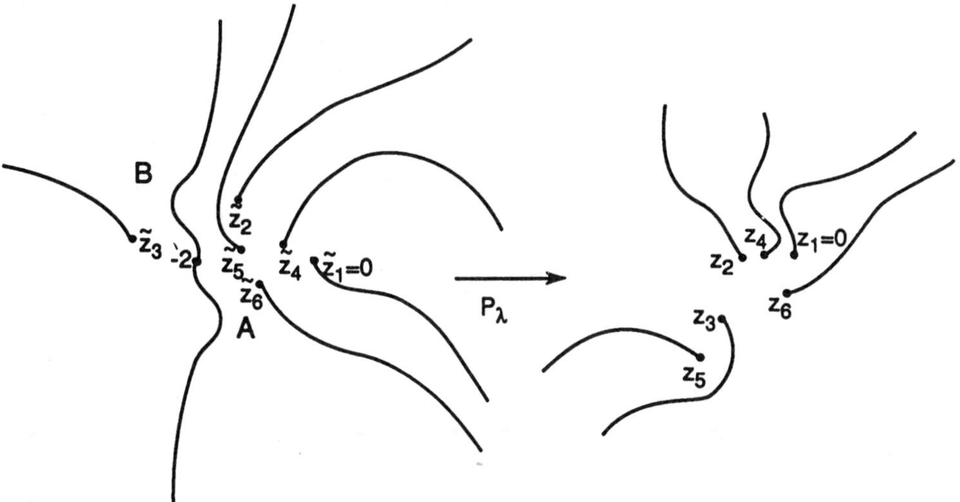

If the angle θ is periodic, so that the kneading sequence contains a symbol $*_i$, the last endpoint $\tilde{z}_n$ is the only inverse image -2 of the critical value $z_{n+1} = z_1 = 0$, and the leg $\tilde{\gamma}_n$ is one of the two inverse images of γ_1 which form the boundary of the two sectors A and B. Using the same circular order as in the definition of the kneading sequence, we can label these two inverse images $*_1$ and $*_2$ and proceed analogously.

The spider space as a complex manifold.

Note that the spider space contains both analytic and combinatorial information: the ratios of the endpoints are analytic functions on the spider space, while the legs supply combinatorial information, since only their homotopy class is defined. This is made precise in Proposition 3.1.

The mapping $\mathcal{S}_\theta^0 \to \mathbb{C}^{n-2}$ given by

$$\varphi \mapsto \begin{bmatrix} \varphi(x_3)/\varphi(x_2) \\ \vdots \\ \varphi(x_n)/\varphi(x_2) \end{bmatrix}$$

induces a mapping $\pi_\theta : \mathcal{S}_\theta \to \mathbb{C}^{n-2}$, the image of which is the subset $U_n \subset \mathbb{C}^{n-2}$ consisting of points whose coordinates are different from $0, 1$ and from each other.

Proposition 3.1. *The mapping $\pi_\theta : \mathcal{S}_\theta \to U_n$ is a universal covering mapping.*

In particular, $\mathcal{S}_\theta$ is a complex manifold of dimension $n - 2$.

Sketch of Proof. Let $\mathbf{w} = (w_3, \ldots, w_n) \in U_n$, and choose open disks $D_3, \ldots, D_n$ around $w_3, \ldots, w_n$ having disjoint closures and not containing 0 or 1, so that

$$\mathbf{D} = D_3 \times D_4 \times \cdots \times D_n \subset U_n.$$

For any $\varphi \in \mathcal{S}_\theta$ with $\pi_\theta(\varphi) = \mathbf{w}$, we need to construct a continuous section s of π_θ over $\mathbf{D}$.

It is easy but cumbersome to check that φ can be modified within its equivalence class so that $\varphi(x_2) = 1$ and so that for each i, the ith leg does not enter D_j for any $j \neq i$, intersects ∂D_i in precisely one point ζ_i and is straight within D_i. If $\mathbf{w}' = (w_3', \ldots, w_n') \in \mathbf{D}$, we will set $s(\mathbf{w}')$ to be the spider with endpoints the w_i', and with legs coinciding with those of φ outside of the D_i, and joining ζ_i to w_i' by the line segment within D_i.

Finally, to see that this covering space is universal, we need to show that $\mathcal{S}_\theta$ is simply connected. Choose a circle $\varphi_t, t \in S^1$ in $\mathcal{S}_\theta$, and a big circle containing all the endpoints of all φ_t. Pull the endpoints back along their legs until they reach the last intersection of the leg with the circle. You then have a collection of points on the circle with the correct circular order; these can be moved into standard position without collisions. $\square$

The knowledgeable reader will see that we have provided an alternative proof that Teichmüller spaces of genus 0 are contractible. This result is known in general as Teichmüller's Theorem.

4. Fixed Points of the Spider Map

The main thing to realize is that if the spider mapping has a fixed point φ, and if we set $\lambda = \varphi(x_2)$, then the polynomial

$$P_\lambda(z) = \lambda \left(1 + \frac{z}{2}\right)^2$$

does have combinatorics which reflect the digits of θ written in base 2. The orbit of the critical point -2 is

$$-2 \mapsto 0 = z_1 = \tilde{z}_1 \mapsto z_2 = \tilde{z}_2 \mapsto z_3 = \tilde{z}_3 \mapsto z_4 \ldots,$$

so it repeats in exactly the same way as the angles $2^j \theta$, i.e., as the digits of θ written in base 2.

Of course, it does rather better than just that: the orbit of the critical point has the same circular order as the points $2^j \theta$ on the circle $\mathbb{R}/\mathbb{Z}$. This requires a bit of amplification: after all, these are just points in $\mathbb{C}$, and don't a priori have a circular order. What does have a circular order are the legs, but they are not obviously related to the polynomial P_λ and its filled-in Julia set K_λ. The connection comes from the following statement:

Proposition 4.1. *Suppose that θ is preperiodic but not periodic under angle doubling, and that the spider map σ_θ has a fixed point φ_θ. Let $\lambda = \varphi_\theta(x_2)$. Then, for every $\theta_j = 2^{j-1}\theta$, the external ray of K_λ at angle θ_j lands at $z_j = P_\lambda^{j-1}(0)$. The union of these rays is (equivalent to) the fixed spider φ_θ.*

The proof of this statement consists of considering the legs not as mere homotopy classes, but as geometric curves. The spider map is still defined, and under iteration, arbitrary legs will approach external rays. The details (of a more general result) can be found in [BFH, Thm I].

We see that a better way of saying that the orbit of the critical point reflects the digits of θ is to say that there is an external ray landing at the critical value, the orbit of which reflects the digits of θ.

Remark. We will see in the next section that something analogous is true even if σ_θ does not have a fixed point in $\mathcal{S}_\theta$.

To give the analogous statement when θ is periodic under angle doubling, we need a bit of terminology.

Let P be a quadratic polynomial with superattracting cycle $z_0, z_1, \ldots, z_{n-1}, z_n = z_0$, where z_0 is critical, and let $U_j \subset \overset{\circ}{K}_\lambda$ be the component containing z_j. There is a unique *internal parametrization* $\psi_j : \overline{\mathbb{D}} \to \overline{U_j}$ conjugating $z \mapsto z^2$ to $P^{\circ n} : \overline{U_j} \to \overline{U_j}$. We call $\psi_j(1)$ the root of U_j: the roots form a repelling cycle of period dividing n.

Proposition 4.2. *Suppose that θ is periodic of period n under angle doubling, and that φ_θ is a fixed point of the spider map σ_θ; let $\lambda = \varphi_\theta(x_2)$. Then the polynomial P_λ has a superattracting cycle of period n, and the external ray of K_λ at angle θ_j lands at the root of U_j. The j-th leg of φ_θ is homotopic rel the critical orbit to the union of the external ray at angle θ_j and the "internal ray" $\psi_j([0,1])$ joining z_j to the root of U_j.*

Remark. As we will see in Section 7, the spider map always has a unique fixed point in the periodic case, but the union of external and internal rays above may (just barely) fail to be the fixed spider. If the period of the roots strictly divides n, each of these points will be used by several legs, so the injectivity condition is violated. This can be cured by an arbitrarily small homotopy: the legs "touch" but do not "cross".

5. Does a Fixed Point of the Spider Mapping exist?

The statement below on the convergence of the spider map is a special case of a deep theorem due to Thurston (see [DH2]). Its proof is quite difficult indeed. We will give a proof when θ is periodic under angle doubling, but even for quadratic spiders, when θ is preperiodic but not periodic under angle doubling, we know of no essential simplification of Thurston's proof.

Theorem 5.1. *For rational θ, let $\varphi \in \mathcal{S}_\theta$, and set*

$$\varphi^{(1)} = \sigma_\theta(\varphi), \; \varphi^{(2)} = \sigma_\theta(\varphi^{(1)}), \ldots .$$

Further, set $z_j^{(n)} = \varphi^{(n)}(x_j)$. Then for each j the sequence

$$z_j, z_j^{(1)}, \ldots, z_j^{(n)}, \ldots$$

converges. Moreover, the ith and jth sequence will have the same limit if and only if the kneading sequence of θ from the ith position on is the same as from the jth position on, i.e., if $k_\theta(2^{i-1}\theta) = k_\theta(2^{j-1}\theta)$.

Notice that we were careful not to say that the sequence of spiders $\varphi^{(n)}$ converges in $\mathcal{S}_\theta$; recall that spiders are required to be injective, and the limits of such sequences may fail to be injective; the second part of the statement says that this occurs exactly if the periodic part of the kneading sequence repeats with a period which strictly divides that of the angle θ.

6. An example of a Thurston Obstruction

In this section we will show with an example how the existence of systems of simple closed curves with certain properties can prevent the spider map from having a fixed point. Our example will show a special case of a Thurston obstruction, which consists of a single simple closed curve. At the end we give as exercises to find Thurston obstructions consisting of two or more curves.

Let us consider again our example $\theta = 9/56$. In the following picture, you see on the right a spider, and on the left its full inverse image, including both the $\tilde{z}_i$, and the other unused inverse images $\tilde{w}_i$. We have also drawn in, on the right, a simple closed curve γ which surrounds z_4, z_5, z_6, the homotopy class of which is completely specified by requiring that it not intersect the legs to the points not surrounded, and that it intersect the legs to the points surrounded exactly once.

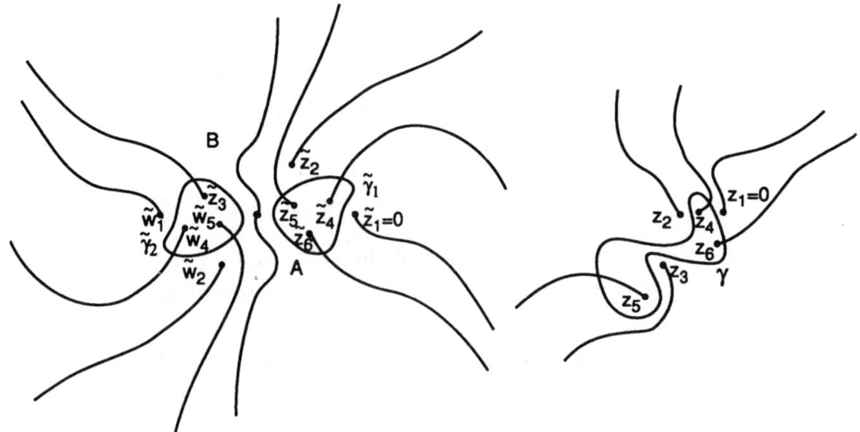

Its complete inverse image $\tilde{\gamma}_1 \cup \tilde{\gamma}_2$ is drawn in on the left. To check that this is really so, note first that γ did not intersect the leg to z_1, so its inverse image will have one component in sector A and one in B. Moreover, all inverse images of z_4, z_5, z_6 must be surrounded by one of the two components, but no other inverse image of any z_i. Inverse images of legs to points not surrounded must not be intersected, however. This specifies the homotopy classes of the components of the inverse images.

The important thing to notice is that $\tilde{\gamma}_1$ is in the same homotopy class as γ.

Now consider the lengths of the curves γ and $\tilde{\gamma}_1$ in the hyperbolic metric of

$$\mathbb{C} - \{z_1, \ldots, z_6\} \quad \text{and} \quad \mathbb{C} - \{\tilde{z}_1, \ldots, \tilde{z}_6\}$$

and denote them by $l_\varphi(\gamma)$ and $l_{\tilde\varphi}(\tilde{\gamma}_1)$.

Lemma 6.1. *We have the inequality*

$$l_{\tilde\varphi}(\tilde{\gamma}_1) < l_\varphi(\gamma).$$

In particular, $\sigma_{9/56}$ cannot have a fixed point in the spider space.

Proof. The mapping P_λ is a covering map, hence an infinitesimal isometry in the Poincaré metric, if considered as a mapping

$$\mathbb{C} - P_\lambda^{-1}\{z_1, \ldots, z_6\} \to \mathbb{C} - \{z_1, \ldots, z_6\}.$$

So in the metric of $\mathbb{C} - P_\lambda^{-1}\{z_1, \ldots, z_6\}$, i.e., in the metric of the complement of all the points drawn, $\tilde{\gamma}_1$ has the same length as γ in $\mathbb{C} - \{z_1, \ldots, z_6\}$.

The inclusion

$$\mathbb{C} - P_\lambda^{-1}\{z_1, \ldots, z_6\} \to \mathbb{C} - \{\tilde{z}_1, \ldots, \tilde{z}_6\}$$

is strictly contracting in the Poincaré metric (like all strict inclusions of Riemann surfaces), so that the length of $\tilde{\gamma}_1$ is strictly smaller than the length of γ.

The statement about the spider map follows: if $\sigma_{9/56}$ had a fixed point, let γ' be the geodesic in the homotopy class of γ on $\mathbb{C} - \{z_1, \ldots, z_6\}$. Then the length of the component $\tilde{\gamma}_1$ of its inverse must be strictly shorter on $\mathbb{C} - \{\tilde{z}_1, \ldots, \tilde{z}_6\}$, and the geodesic $\tilde{\gamma}_1'$ in the homotopy class of $\tilde{\gamma}_1$ on $\mathbb{C} - \{\tilde{z}_1, \ldots, \tilde{z}_6\}$ is shorter yet. But there is only one geodesic in a given homotopy class, so $\tilde{\gamma}_1' = \gamma'$. This is a contradiction. $\square$

The reader should observe that the existence of the curve γ, for which one component of the inverse image is homotopic to the curve itself, has everything to do with the fact that the 9/56-itineraries of $2^{j-1}9/56$ coincide for $j = 4, 5, 6$. A similar construction will be possible whenever the symbols in the kneading sequence repeat with period strictly smaller than the period of the angle itself.

Exercise. Show that for the angle $\theta = 27/60$, there is a system of two simple closed curves in the complement of the endpoints, such that one component of the inverse image of each is homotopic to the other. Why does this prevent the map $\sigma_{27/60}$ from having a fixed point?

Exercise. Find an angle θ such that there is a system of 3 curves behaving as above.

7. A Proof of Convergence when θ is Periodic

In this section we will prove the convergence of the spider algorithm in the case when θ is periodic. This is a special case of Thurston's theorem, but it seems worthwhile to give the proof anyway, as it is really very much simpler than the general case.

The proof consists of two parts: a compactness statement, which replaces a much more difficult argument, and a contraction statement, which is somewhat simpler than the more general proof.

Before starting, we will slightly modify our definition of S_θ: we will consider only spiders φ satisfying $\varphi(x_n) = -2$. Note that any spider φ with the previous definition is equivalent to a unique one satisfying this condition, namely

$$\frac{-2}{\varphi(x_n)}\varphi,$$

so this requirement does not change the spider space. Moreover, the image of the spider map consists entirely of spiders fulfilling this requirement. We will freely write $x_0 = x_n$, keeping in mind that indices are only defined modulo n.

The compactness statement.

Our compactness statement will say that a certain subset of $\mathcal{S}_\theta$, in which the endpoints are bounded away from each other and infinity, is invariant under the spider map.

The image of this subset under π_θ is a compact subset of U_n, and it follows, since $\varphi(x_n) = -2$, that the *endpoints* z_j $(j = 1,\ldots,n)$ belong to a compact subset of $\mathbb{C}^n - \Delta$, where Δ is the set where any two coordinates are equal. The set of spiders belonging to this invariant subset is an infinite cover of the space of endpoints and is *not* compact.

The statement is a bit convoluted; the proof should explain why.

Proposition 7.1. *Choose ε satisfying $0 < \varepsilon < 4^{2-n}$ and let $\mathcal{S}_{\theta,\varepsilon}$ be the set of θ-spiders φ such that the discs of radius $4^{j-n}\varepsilon/|z_2|$ around z_j and the disc of radius $\varepsilon/|z_2|$ around $z_0 = z_n = -2$ are disjoint, and all z_j are contained in a disc around -2 of radius $8/|z_2|$. Then $\mathcal{S}_{\theta,\varepsilon}$ is invariant under σ_θ.*

Both statement and proof of this proposition will become much simpler if we choose the right coordinate system. The polynomial $P_\lambda(z) = \lambda(1 + z/2)^2$ is conjugate to $p_c(w) = w^2 + c$ by an affine map ψ if and only if $c = \lambda/2$ (and $\lambda \neq 0$). Specifically, for $w = \psi(z) = (z + 2)\lambda/4$, we have $p_c = \psi \circ P_\lambda \circ \psi^{-1}$. The critical point is now $w_0 = \psi(-2) = 0$ and the critical value is $w_1 = \psi(0) = c$.

Let $\mathbb{D}_2$ be the disc around the origin of radius 2. Then it is easy to verify that a spider φ is in $\mathcal{S}_{\theta,\varepsilon}$ if and only if the points $w_j = \psi(z_j)$ are contained in $\mathbb{D}_2$ and are surrounded by disjoint discs of radius $r_j = 4^{j-1-n}\varepsilon$, for $j = 1, \ldots, n$ (remember $\lambda = z_2$).

The result now follows from applying ψ^{-1} to the discs provided by the following lemma. Note that this lemma does not require the right choice of inverse images to be taken.

Lemma 7.2. *Choose ε as above and suppose that $w_0 = 0$, $w_1 = c$, $\ldots$, w_{n-1}, $w_n = w_0$ are contained in $\mathbb{D}_2$ such that the discs of radius $r_j = 4^{j-1-n}\varepsilon$ around w_j are disjoint for $j = 1, \ldots, n$. For each $j = 0, \ldots, n-1$, let $\tilde{w}_j$ be an element of $p_c^{-1}(w_{j+1})$ (which forces $\tilde{w}_0 = 0$), and set $\tilde{w}_n = 0$ also. Then all these $\tilde{w}_j$ will be contained in $\mathbb{D}_2$, and the discs of radius r_j around them will again be disjoint for $j = 1, \ldots, n$.*

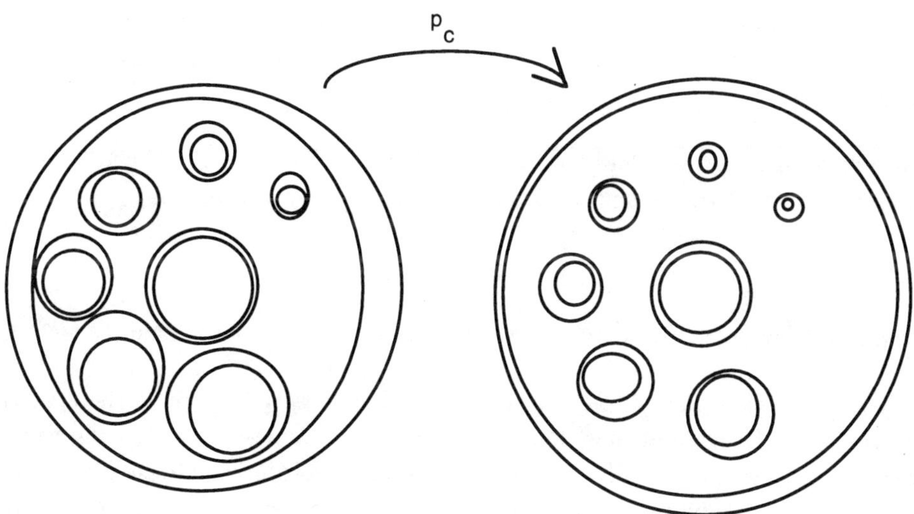

Proof of Lemma 7.2. In order to show that each inverse image $\tilde{w}_j = \sqrt{w_{j+1} - c}$ is contained in $\mathbb{D}_2$, it suffices to estimate $|w_{j+1} - c| < |w_{j+1}| + |c| < 4$.

To exclude the possibility that points get too close together, we first note that the inverse image of the disc around $w_1 = c$ of radius $r_1 = 4^{-n}\varepsilon$ is a disc around $\tilde{w}_0 = 0$ of radius $4^{-n/2}\sqrt{\varepsilon}$ and hence contains the disc $D_{r_0}(0)$ of radius $r_0 = r_n = 4^{-1}\varepsilon < 4^{-1}4^{(2-n)/2}\sqrt{\varepsilon} = 4^{-n/2}\sqrt{\varepsilon}$.

Note that $|p'_c| < 4$ on $\mathbb{D}_2$. It follows that if $\tilde{w} \in \mathbb{D}_2$ is contained in a disc around some $\tilde{w}_j$ of radius r_j ($1 \le j \le n-1$), then $w := p_c(\tilde{w})$ is contained in a disc around $w_{j+1} = p_c(\tilde{w}_j)$ of radius $4r_j = r_{j+1}$.

Now, if $D_{r_i}(\tilde{w}_i)$ and $D_{r_j}(\tilde{w}_j)$ had non-empty intersection, we could choose a point $\tilde{w}$ in their intersection which was also contained in $\mathbb{D}_2$. But the image point $w = p_c(\tilde{w})$ would then be contained in the given discs around the image points $p_c(\tilde{w}_i)$ and $p_c(\tilde{w}_j)$, contradicting the hypotheses. $\square$

The contraction statement.

We wish to prove that the spider map is strictly contracting. This requires a metric on the space of spiders, and defining it is not quite straightforward.

We will define an infinitesimal metric, which will be given by a norm (which is *not* an inner product) on the tangent spaces. This kind of structure is called a Finsler metric; such structures carry a great deal of geometry. Many of the deepest results about Teichmüller spaces are obtained by looking really carefully at the unit balls for such metrics.

Let θ be periodic of period n. The tangent space to $\mathcal{S}_\theta$ at a spider φ with endpoints $Z_\varphi = \{z_0 = -2, z_1 = 0, z_2, \ldots, z_{n-1}\}$ is the space

$$T_\varphi \mathcal{S}_\theta = \{\underline{\xi}\} := \left\{ \left(\vec{\xi}_2, \ldots \vec{\xi}_{n-1} \right) \middle| \vec{\xi}_j = \xi_j \, \partial/\partial z \in T_{z_j}\overline{\mathbb{C}} \right\}.$$

This follows from the fact that the mapping which associates to a spider its endpoints is a covering map, so the tangent space to the spider space is simply the tangent space to $\mathbb{C}^{n-2}$.

Remark. Although we will be working on $\mathbb{C}$, where there is a global chart, we will keep tangent vectors distinct from complex numbers, by writing

$$\vec{\xi} = \xi \frac{\partial}{\partial z}.$$

Confusing numbers and tangent vectors can lead to conceptual difficulties.

The cotangent space $T^*_\varphi \mathcal{S}_\theta$ also has a very nice description: it is the space of polynomials of degree at most $n-3$. This is not an honest way of saying things: the cotangent space is really the space Q_φ of meromorphic quadratic differentials on $\overline{\mathbb{C}}$ with poles only on $Z_\varphi \cup \{\infty\}$ and all poles at most simple. All such quadratic differentials can be written

$$q = \frac{p(z)}{\prod_{j=0}^{n-1}(z - z_j)} dz^2 \tag{1}$$

with p a polynomial of degree at most $n-3$. The reason for this restriction is that the denominator has a zero of order n at ∞, the quadratic differential dz^2 has a pole of order 4, so that if $\deg p \le n-3$, then q has at most a simple pole at ∞.

The fact that Q_φ is the cotangent space is really a special case of the Serre Duality Theorem. In this case, the duality can be made quite explicit.

Proposition 7.3. *The pairing*

$$\langle \underline{\xi}, q \rangle = \sum_{j=2}^{n-1} \mathrm{Res}_{z_j} \vec{\xi}_j q = \sum_{j=2}^{n-1} \xi_j \frac{p(z_j)}{\prod_{i \neq j}(z_j - z_i)}$$

induces a duality between $T_\varphi S_\theta$ *and* Q_φ.

Remark. The product of a quadratic differential and a vector field is naturally a differential 1-form, as is suggested by the equation

$$\left(q(z)dz^2\right) \left(\xi(z)\frac{\partial}{\partial z}\right) = \xi(z)q(z)dz.$$

Thus the residue in the statement is really the residue of a 1-form; this is the kind of residue which is studied in elementary complex analysis.

Proof. It is clear that we have written a pairing, which induces a mapping of each space into the dual of the other. Think of it as a mapping

$$T_\varphi S_\theta \to (Q_\varphi)^* \ ;$$

we will show it is injective. If $\underline{\xi} = (\vec{\xi}_2, \dots, \vec{\xi}_{n-1}) \neq 0$, there exists j such that $\xi_j \neq 0$. Then $\underline{\xi}$ pairs non-trivially with the quadratic differential

$$\frac{dz^2}{z(z+2)(z-z_j)}$$

which is of the form (1) with

$$p(z) = \prod_{i=2, \ i\neq j}^{n-1} (z - z_i).$$

Since $T_\varphi S_\theta$ and Q_φ both have dimension $n-2$, this proves the Proposition. $\square$

If $q = q(z)dz^2$ is a quadratic differential, then $|q| := |q(z)|dx\,dy$ is a positive measure in a natural way. Moreover, simple poles are integrable:

$$\int_{\mathbb{D}} \left| \frac{dz^2}{z} \right| = \int_{\mathbb{D}} \frac{1}{|z|} dx\,dy = 2\pi$$

is finite. Thus we can define the L^1 norm on Q_φ by the formula

$$\|q\| = \int_{\mathbb{C}} |q|.$$

Exercise. Check directly that the condition $\deg p \leq n - 3$ is equivalent to

$$\int_{\mathbb{C}} \left| \frac{p(z)}{\prod_{j=0}^{n-1}(z - z_j)} \right| dx\, dy < \infty.$$

Note that the condition φ injective, which keeps the z_j distinct, is essential for this statement to be true: double poles are *not* integrable.

We are finally in a position to define our metric on the spider space: it is the metric given by the norm on the tangent spaces dual to the L^1 norm above on the cotangent spaces. Recall that if E is a Banach space (in our case finite dimensional) with norm $\| \ \|_E$, then the dual space E^* (the space of continuous linear functionals $E \to \mathbb{C}$) naturally carries the norm

$$\|\alpha\|_{E^*} = \sup_{x \in E - \{0\}} \frac{|\alpha(x)|}{\|x\|_E}.$$

Proposition 7.4. *The spider mapping is strictly contracting, i.e.,*

$$\|d_\varphi \sigma_\theta\| < 1$$

for all θ-spiders φ.

Moreover, the norm of the derivative depends only on the endpoints of φ and $\tilde{\varphi}$, and not on the combinatorial information contained in the legs.

We first need to define the operator

$$(P_\lambda)_* : Q_{\tilde{\varphi}} \to Q_\varphi.$$

Choose a point $z \in \mathbb{C}$. Suppose $z \neq 0$, and choose a simply connected neighborhood U of z which does not include 0, and let ζ be a local coordinate in U. Since U contains no critical value of P_λ, $P_\lambda^{-1}(U)$ consists of two components U_1, U_2, in which the lifts ζ_1, ζ_2 of ζ are local coordinates. Any $q \in Q_{\tilde{\varphi}}$ (in fact, any quadratic differential in the "domain" Riemann surface) can be written $q_i(\zeta_i)d\zeta_i^2$ on U_i. We define

$$(P_\lambda)_* q = (q_1(\zeta) + q_2(\zeta))d\zeta^2 \tag{2}$$

on U. This defines $(P_\lambda)_* q$ everywhere but at the critical value 0. It is clear that $(P_\lambda)_* q$ is holomorphic, except for possible simple poles at the images of the poles of

q under P_λ, and at 0 where *a priori* we only know that it has an isolated singularity. Lemma 7.5 implies that even at 0 it has at worst a simple pole.

Now the proposition follows from the following two lemmas.

Lemma 7.5. *The mapping*

$$(P_\lambda)_* : Q_{\tilde\varphi} \to Q_\varphi.$$

satisfies $\|(P_\lambda)_*\| < 1.$

Lemma 7.6. *The mapping*

$$(P_\lambda)_* : Q_{\tilde\varphi} \to Q_\varphi.$$

is the transpose (also known as adjoint) of $d_\varphi \sigma_\theta.$

Proof of Lemma 7.5. Use the notation of (2) and before. For every U which does not contain the critical value, we have

$$\int_U |(P_\lambda)_* q| \leq \int_U |q_1| + \int_U |q_2| = \int_{P_\lambda^{-1}(U)} |q|$$

by the triangle inequality; it follows easily that $\|(P_\lambda)_*\| \leq 1$. If $\|(P_\lambda)_*\| = 1$, there exists $q \in Q_{\tilde\varphi}$ such that

$$\|(P_\lambda)_* q\| = \|q\|, \tag{3}$$

since the spaces are finite dimensional. Where can the poles of such a q be? Every z_j with $j \neq 1$ has two inverse images, $\tilde z_{j-1}$ and another point $\tilde w_{j-1}$ where q cannot have a pole. Choose a neighborhood U of z_j for $j \neq 1$; equality (3) says that nothing is lost in the triangle inequality, i.e. that $q_1 = Cq_2$ for some $C \geq 0$. But one of the U_i contains $\tilde w_{j-1}$, hence q_1 has no pole there, and so q has no pole at $\tilde z_{j-1}$.

The upshot of this analysis is that q satisfying equation (3) can have poles only at -2 and ∞; but a quadratic differential always has at least 4 poles. Thus the case $\|(P_\lambda)_*\| = 1$ is excluded. $\square$

Proof of Lemma 7.6. It is not hard to differentiate σ_θ; if you find the formulas cumbersome, look for an explanation below. Suppose $\sigma_\theta(\varphi) = \tilde\varphi$; the endpoints of φ and $\tilde\varphi$ are called z_j and $\tilde z_j$, respectively, for $j = 0, \ldots, n-1$ as usual. If $d_\varphi \sigma_\theta(\underline\xi) = \underline{\tilde\xi}$, then differentiating the relation

$$P_{z_2}(\tilde z_j) = z_{j+1}$$

yields

$$\xi_2 \left(1 + \frac{\tilde{z}_j}{2}\right)^2 + z_2 \left(1 + \frac{\tilde{z}_j}{2}\right) \tilde{\xi}_j = \xi_{j+1}.$$

Solving for $\tilde{\xi}_j$, we find

$$\tilde{\xi}_j = \left(\frac{1}{z_2(1 + \tilde{z}_j/2)}\right) \xi_{j+1} - \left(\frac{1 + \tilde{z}_j/2}{z_2}\right) \xi_2. \tag{4}$$

Now we define

$$\mu_j := \left(\frac{1}{z_2(1 + \tilde{z}_j/2)}\right) \xi_{j+1}$$

so that

$$\vec{\mu}_j := \mu_j \frac{\partial}{\partial z} = \left(d_{\tilde{z}_j} P_{z_2}\right)^{-1} \vec{\xi}_{j+1}.$$

Moreover, let

$$\vec{\mu} := \mu(\tilde{z}) \frac{\partial}{\partial \tilde{z}} = \left(\frac{1 + \tilde{z}/2}{z_2}\right) \xi_2 \frac{\partial}{\partial \tilde{z}}$$

be the global vector field on $\overline{\mathbb{C}}$ which vanishes at -2 and ∞, and the value of which at $0 = \tilde{z}_1$ is

$$\left(d_{\tilde{z}_j} P_{z_2}\right)^{-1} \xi_2 \frac{\partial}{\partial \tilde{z}} = \left(\frac{1}{z_2}\right) \xi_2 \frac{\partial}{\partial \tilde{z}}.$$

Using this notation, equation (4) becomes

$$\tilde{\xi}_j = \mu_j - \mu(\tilde{z}_j).$$

A word of explanation of what these formulae mean: one might well expect that the derivative of σ_θ would simply take the tangent vectors $\vec{\xi}_{j+1}$, and lift each one to a vector at $\tilde{z}_j$ by the derivative of P_λ. This construction yields the tangent vectors $\vec{\mu}_j$, but it isn't quite the required derivative. In particular, $\vec{\mu}_1$ may well fail to vanish, even though $\tilde{z}_1 = 0$ is constant. The correct derivative is given by subtracting from each $\vec{\mu}_j$ the value at z_j of the unique global vector field which vanishes at -2 and ∞, and which is equal to $\vec{\mu}_1$ at 0.

The formula we need to verify is that

$$\langle \tilde{\underline{\xi}}, q \rangle = \langle \underline{\xi}, (P_\lambda)_* q \rangle$$

for any $q \in Q_{\tilde{\varphi}}$ and any $\underline{\xi} \in T_\varphi \mathcal{S}_\theta$.

By the definition of the pairing, we have

$$\langle \tilde{\underline{\xi}}, q \rangle = \sum_{j=2}^{n-1} \mathrm{Res}_{\tilde{z}_j} \vec{\tilde{\xi}}_j q$$

$$= \sum_{j=1}^{n-1} \mathrm{Res}_{\tilde{z}_j} \vec{\mu}_j q - \sum_{j=1}^{n-1} \mathrm{Res}_{\tilde{z}_j} \vec{\mu}(z_j) q. \tag{5}$$

Note that the terms corresponding to $j = 1$ in the two sums cancel. The last sum in equation (5) vanishes by the Cauchy integral formula, since $\bar{\mu}q$ is a global 1-form on the Riemann sphere, with poles only at the $\tilde{z}_j$ for $j = 1, \ldots, n-1$ (and not at -2 and ∞, since $\bar{\mu}$ vanishes there). Moreover, $\mu_{n-1} = 0$ since it is the pull-back of the tangent vector to $z_0 = -2$ which is fixed. So altogether we get

$$\langle \tilde{\underline{\xi}}, q \rangle = \sum_{j=1}^{n-2} \operatorname{Res}_{\tilde{z}_j} \bar{\mu}_j q \ .$$

On the other hand, by the definition of the push-forward, we have

$$\begin{aligned}
\langle \underline{\xi}, (P_\lambda)_* q \rangle &= \sum_{j=2}^{n-1} \operatorname{Res}_{z_j} \vec{\xi}_j \, (P_\lambda)_* \, q \\
&= \sum_{j=1}^{n-2} \operatorname{Res}_{\tilde{z}_j} \vec{\mu}_j q + \sum_{j=1}^{n-2} \operatorname{Res}_{\tilde{w}_j} \vec{\nu}_j q
\end{aligned} \tag{6}$$

where $\tilde{w}_j$ is the inverse image of z_{j+1} which is not $\tilde{z}_j$, and

$$\nu_j = \left(d_{\tilde{w}_j} P_{z_2} \right)^{-1} \xi_{j+1}$$

is the lift of ξ_{j+1} there. The switch of indices from $2, \ldots, n-1$ to $1, \ldots, n-2$ comes from the labeling of inverse images. The second sum in equation (6) vanishes because q does not have poles at the $\tilde{w}_j$. Putting together equations (5) and (6) (and remembering that the last sums in both of them vanish) proves Lemma 7.6.

Proof of Thurston's Theorem. We are now in a position to prove the main result:

Theorem 7.7. *If θ is periodic under angle doubling, there is a unique fixed point φ_θ of σ_θ in $\mathcal{S}_\theta$, and all of $\mathcal{S}_\theta$ is attracted to it under interation of σ_θ.*

Proof. Recall the space $\mathcal{S}_{\theta,\varepsilon}$ from Proposition 7.1. We will require the following statement.

Lemma 7.8. *For any two spiders $\varphi_0, \varphi_1 \in \mathcal{S}_{\theta,\varepsilon}$, there exists $\delta > 0$ and a path γ joining φ_0, φ_1 contained in $\mathcal{S}_{\theta,\delta}$.*

In Proposition 3.1, we showed much more than connectivity of the spider space, we showed contractibility. Recall that this required drawing a large circle, and pulling the legs back to their last intersection with this circle. A similar proof gives Lemma 7.8: move to the w-plane, as in the proof of Lemma 7.2, and repeat the argument, but with the circle of radius 2. You then will obtain a path connecting the two spiders, with all endpoints staying in the disc of radius 2. The endpoints will stay some distance apart, and it is easy to choose δ in terms of that distance.

To continue the proof of Thurston's theorem, choose $\varphi_0 \in \mathcal{S}_{\theta,\varepsilon}$, set $\varphi_1 = \sigma_\theta(\varphi_0)$, and find δ and a path γ_0 connecting φ_0 to φ_1 in $\mathcal{S}_{\theta,\delta}$, which we may suppose continuously differentiable, hence of finite length L. Set $\varphi_m = \sigma_\theta^m(\varphi_0)$, and $\gamma_m = \sigma_\theta^m(\gamma_0)$ for $m \geq 2$. All of these φ_m and γ_m are contained in $\mathcal{S}_{\theta,\delta}$, so Proposition 7.4 says that there is a constant $C_\delta < 1$ such that the length of γ_m is at most $C_\delta^m L$. In particular, the sequence φ_m is a Cauchy sequence.

This is not quite enough to know that it converges: we would need to know that $\mathcal{S}_\theta$ is complete, which is true but rather harder to see than anything we have used so far.

Instead, we argue as follows. The endpoints $z_{j,m}$ of φ_m certainly converge, say to y_j, where the y_j are distinct, and with distances bounded below since all φ_m belong to $\mathcal{S}_{\theta,\delta}$ for a fixed δ. It is now easy to see that if the φ_m do not converge, there exist distinct θ-spiders with these same endpoints y_j which are arbitrarily close together. But to move from one δ-spider to another with the same endpoints, at least one endpoint will have to leave its disc. The length of such a path is certainly bounded below by a positive constant which depends only on the y_j. $\square$

8. Extensions to Higher Degrees

Everything we have done extends to the family of polynomials $P_{d,\lambda}(z) = \lambda(1 + z/d)^d$. The standard θ-spider is now defined using the mapping $\mathbb{T} \to \mathbb{T}$ given by $t \mapsto dt$, rather than angle-doubling. The kneading sequence is now a sequence of sectors limited by the d angles $\theta/d, (\theta + 1)/d, \ldots, (\theta + d - 1)/d$.

We can define the space $\mathcal{S}_{d,\theta}$ exactly as for degree 2, and the spider map

$$\sigma_{d,\theta} : \mathcal{S}_{d,\theta} \to \mathcal{S}_{d,\theta}$$

is now easy to define, once we realize that the inverse image of the leg to $z_1 = 0$ consists of d rays landing at $-d$, cutting the plane into d sectors, which each contain exactly one inverse image of every point not on the leg of z_1.

We can even extend the spider mapping to the family of exponentials when the "angle" is preperiodic; the d-th root which appears in $\sigma_{d,\theta}$ becomes a logarithm. Thurston's proof for the convergence of the algorithm does not carry over to exponentials.

The polynomials $P_{d,\lambda}(z) = \lambda(1 + z/d)^d$ are special: they have only one critical point. The spider algorithm can be adapted to polynomials with arbitrarily many critical points, all of the orbits of which are finite. The underlying mathematics is done in [BFH] for the preperiodic case, and in [P] in general. To give details would take us too far afield.

The "elementary" proof of convergence is a bit delicate to generalize to this more general situation. The proof of Proposition 7.1 will generalize to arbitrary polynomials for which every critical point lands on a cycle containing a critical point. This is a substantial simplification of Thurston's proof in that case. The statement of Proposition 7.4 is true in full generality, but the proof we have given can only be extended to rational functions with at most three critical values.

9. A Comment on Implementation

Any one thinking of implementing the spider algorithm should pause when notions like "continuous curve", and "choose the inverse image on this side of the continuous curve ... " appear. It is notoriously difficult to simulate the continuous on the computer; in particular, the idea of encoding a leg by a sequence of pixels is a bit repelling.

Fortunately, it is not necessary. In fact, the algorithm can be encoded in a remarkably clean way.

We will suppose that spiders are encoded as endpoints $z_1 = 0, z_2, \ldots, z_n$ (floating-point complex numbers) and legs (polygonal lines in the plane, going from z_j to a circle of sufficiently large radius).

There are two essential steps in the spider algorithm: computing the inverse images $\tilde{z}_j$ in the appropriate sectors, and computing the new legs. Let us take these in order.

Computing the New Endpoints. Among the inverse images $P_{d,z_2}^{-1}(z_{j+1})$ of z_{j+1}, let $\tilde{w}_j$ be the one closest to 0. What sector does it belong to? If we can answer that, then we can find the inverse image which belongs to the required sector. It would be quite hard to answer this question using the description given in Section 3, because that used the inverse images of the leg leading to z_1, and we don't have these inverse images available.

We claim that $\tilde{w}_j$ is in the sector numbered (mod d) by the algebraic intersection number $I(\gamma_1, [z_2, z_{j+1}])$ of the leg γ_1 to z_1 and of the segment $[z_2, z_{j+1}]$, because the lift of this segment starting at 0 leads to $\tilde{w}_j$. More precisely, orient γ_1 so that it goes from z_1 to ∞, and orient $[z_2, z_{j+1}]$ so that it goes from z_2 to z_{j+1}. Then this algebraic intersection number is the sum over all intersections of the numbers $+1$ if the leg crosses the segment from left to right and -1 if it crosses from right to left.

The number $I(\gamma_1, [z_2, z_{j+1}])$ is the sum, over all the oriented line segments s_j of the leg, of the intersections $s_j \cap [z_2, z_{j+1}]$. Thus we see that for this part of the program, the central subroutine is the one which, given two line segments defined by their endpoints, tells whether they intersect, and if so, with which orientation. This subroutine uses about half the computer time, and it pays to do it efficiently.

Lifting the Legs. The inverse image of a polygonal leg is a curve formed of segments of curves which look like hyperbolas (and are hyperbolas for $d = 2$). We need to replace these by polygonal curves in the same homotopy classes.

It may not be correct to simply join the inverse images of the vertices. If an endpoint $\tilde{z}_j$ (which we have already found) is inside the convex hull of a segment of "hyperbola", then replacing the segment of "hyperbola" by the segment of straight line would change the homotopy class of the leg.

So one must discover whether that is the case or not; the computer uses the other half of its time here. It requires three checks:

(1) that $\tilde{z}_j$ is on the same side as $-d$ of the line,

(2) that it is inside the "hyperbola", which means that z_{j+1} and 0 are on opposite sides of the line carrying the segment of which we are computing an inverse image, and finally

(3) that seen from $-d$, $\tilde{z}_j$ and the endpoints of the segment of hyperbola make an angle smaller than π/d.

Each one of these checks is verifying one simple inequality (and it really pays to encode it economically).

Of course, one needs to decide what to do if a segment of hyperbola does "enclose" an endpoint, which fortunately doesn't happen often. Recursively add points cutting the hyperbola into smaller pieces until they don't. This may mean that a leg is encoded by a lot of segments, and it is best to prune before continuing.

Bibliography.

[BFH] B. Bielefeld, Y. Fisher, J. H. Hubbard, *The classification of critically preperiodic polynomials as dynamical systems*, Jour. A.M.S., Vol. 5, No. 4 (1992).

[DH1] A. Douady, J. H. Hubbard, *Etude dynamique des polynômes complexes*, Université de Paris-Sud, 1984/85.

[DH2] A. Douady, J. H. Hubbard, *A proof of Thurston's topological characterization of rational functions*, Acta Math. 171 (1993).

[DGHS] R. Devaney, L. Goldberg, J. H. Hubbard, D. Schleicher, *A dynamical approximation to the exponential map by polynomials*, to appear.

[MT] J. Milnor, W. Thurston, *On iterated maps on the interval*, Lecture Notes in Mathematics 1342 (1988), 465-563.

[P] A. Poirier, *On postcritically finite polynomials*, parts I and II. Stony Brook Preprint, 5 and 7 (1993).

[S] D. Schleicher, *Rational external rays of the Mandelbrot set*. In: Thesis, Cornell University (1994).

[W] E. B. White, *Charlotte's Web* (pictures by Garth Williams), Harper Trophy (1952).

Proceedings of Symposia in Applied Mathematics
Volume **49**, 1994

Complex Dynamics and Entire Functions

Robert L. Devaney*

Our goal in this paper is to describe some of the similarities and distinctions between the iteration of rational maps and that of entire transcendental functions. The main differences arise because of the point at ∞. For polynomials, this point is a superattracting fixed point. For rational maps, ∞ is dynamically the same as any other point on the Riemann sphere. But for entire transcendental functions, ∞ is an essential singularity. By the Picard theorem, any neighborhood of ∞ is mapped by an entire function over the entire plane, missing at most one point, and taking any other value infinitely often. Thus the point at ∞ injects a tremendous amount of "hyperbolicity" into the dynamics. This results in substantial differences in both the dynamics on and the geometry of the Julia sets for these types of maps, as we shall see below. It is a pleasure to thank Tom Scavo for many helpful comments on an earlier draft of this paper.

1. Singular values. As we have seen in previous chapters in this book, the dynamics of polynomials and rational maps is determined in large measure by the fate of the orbits of critical values. For entire functions, the set of "singular" orbits must be extended, as shown by the following example.

Example. Let $E_\lambda(z) = \lambda e^z$ with $\lambda \in \mathbf{R}$. Suppose λ is small ($\lambda < 1/e$

---

* 1980 *Mathematics Subject Classification (1985 Revision)*. Primary 58F13; Secondary 58F14. Partially supported by National Science Foundation Grant DMS-93-02986

suffices). Then the graph of E_λ shows that E_λ has an attracting fixed point. See Figure 1. But E_λ has no critical points whatsoever.

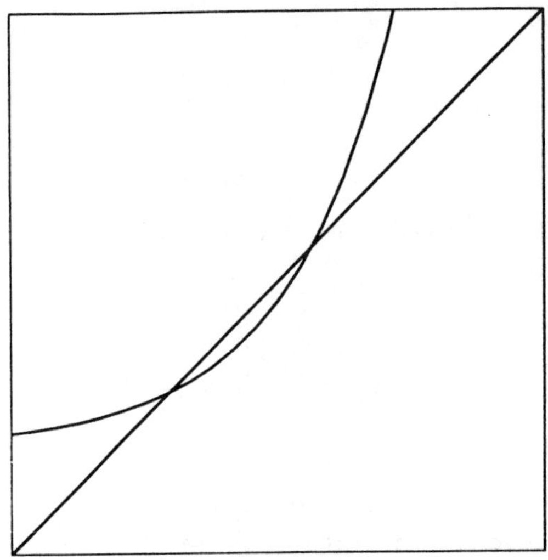

Fig. 1 $E_\lambda(x) = \lambda e^x$ has an attracting fixed point for λ small.

In the above example, the role of the critical value is played by 0, which is an omitted value for E_λ. Technically, 0 is an *asymptotic value*.

Definition. Suppose $F: \mathbf{C} \to \mathbf{C}$ is an entire transcendental function. A point $w \in \overline{\mathbf{C}}$ is an asymptotic value for F if there is a continuous curve $\gamma(t)$ satisfying

$$\lim_{t \to \infty} \gamma(t) = \infty$$

and

$$\lim_{t \to \infty} F(\gamma(t)) = w.$$

Note that any curve which tends to ∞ such that Re $z \to -\infty$ is such a curve γ for the exponential function. Hence 0 is an asymptotic value for E_λ.

One major difference between entire functions and rational maps is the following fact.

Theorem. *Suppose E is an entire transcendental function and z_0 lies on an attracting or parabolic cycle for E. Then the orbit of at least one critical point or asymptotic value is attracted to the orbit of z_0.*

In general, analogous results concerning the role of critical value as discussed in Keen's article hold for entire functions, provided we augment the set of singular orbits to include asymptotic as well as critical value. Note that an entire function cannot have Herman rings. This holds for the same reason that polynomials have no such stable regions.

2. Stable regions. For entire functions, the classification of stable domains must be expanded to include two more types. Besides attracting basins, superattracting basins, parabolic domains, and Siegel disks, entire functions may admit *wandering domains* or *Baker domains*.

Example. Consider the function $F_\lambda(z) = z + \lambda \sin z$ where λ is real. Note that F_λ has critical points where $\cos z = -1/\lambda$. If $\lambda > 1$ there are infinitely many critical points on the real line for F_λ. By a judicious choice of λ we may arrange that each critical point z_0 is mapped to $z_0 \pm 2\pi$, i.e., to another critical point. Thus there are only two critical orbits. See Figure 2. As a consequence, any sufficiently small neighborhood of a critical point lies in the stable set since successive iterates contract these regions and all orbits tend uniformly to ∞ in these neighborhoods. As we will see below, the vertical lines $x = k\pi$ lie in the Julia set of F_λ for $k \in \mathbf{Z}$, so it follows that each point on a critical orbit lies in a distinct component of the stable set. These components are called wandering domains since they are not eventually periodic.

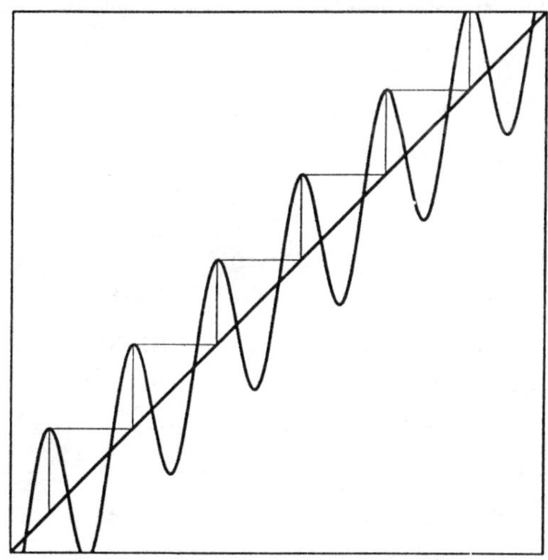

Fig. 2 The graph of $F_\lambda(x) = x + \lambda \sin x$.

Example. The following example is due to Fatou [F]. Consider $F(z) = z + 1 + e^{-z}$. Observe that F has critical points at $z = 2k\pi i$ for $k \in \mathbf{Z}$. If Re $z > 0$, the orbit of z under F tends to ∞. This follows from the observation that if $\operatorname{Re} z > 0$, then

$$\operatorname{Re} F(z) = \operatorname{Re} z + 1 + \operatorname{Re} e^{-z} > \operatorname{Re} z.$$

As a consequence, all orbits in the right-half plane lie in a stable domain consisting of points whose orbits tend to the essential singularity. In particular, the critical values $2 + 2k\pi i$ all tend to ∞. While this set resembles a parabolic domain (with fixed point on the boundary of the domain), the dynamics here are quite different since the "fixed point" at ∞ is really an essential singularity. Such a component of the stable set is called a Baker domain or domain at infinity.

Note that, in each of the previous two examples, the wandering domain or domain at ∞ contains infinitely many critical points. This motivates the following definition.

Definition. An entire transcendental function is critically finite if it has only finitely many asymptotic or critical values.

For example, the entire functions $\lambda \sin z$ and $\lambda \cos z$ are critically finite since these maps have only two critical values and no (finite) asymptotic values. On the other hand, the families $z + \lambda \sin z$ and $z + \lambda \cos z$ are not critically finite.

Critically finite entire maps form a class of entire functions whose dynamics prove to be most tractable. The following result is due to Goldberg and Keen [GK] and Eremenko and Lyubich [EL].

Theorem. *Let S be the class of critically finite entire transcendental functions. If $E \in S$, then E has no wandering domains and no domains at ∞.*

For functions which are not contained in the class S, the possible structures for wandering domains are still not completely catalogued. Baker [Ba] has described some multiply connected wandering domains and Eremenko and Lyubich [EL] have shown the existence of oscillatory wandering domains which accumulate both at ∞ and at finite points in the plane.

The above theorem has as a consequence the fact that Sullivan's classification of stable domains (attracting and superattracting basins, parabolic basins, and Siegel disks) holds for critically finite maps. Each such type of stable domain involves the attraction or accumulation of at least one singular orbit (orbit of a critical point or asymptotic value). This often allows us to eliminate stable domains altogether, provided we know the behavior of all singular orbits.

3. The Julia Set. The Julia set $J(E)$ for an entire function E is defined just as in the rational map case, as the set of points at which the family of iterates of E fails to be a normal family. Fatou [F] showed that periodic points are dense in the Julia set of an entire map, and Baker [Ba] later extended this result to show that $J(E)$ is the closure of the set of *repelling* periodic points. We will see below a third characterization of the Julia set of critically finite entire functions, namely as the closure of the set of points whose orbits tend to ∞. This is in marked contrast to the polynomial case, where orbits tending to ∞ always lie in the stable set.

Entire functions often have Julia sets which cover the entire plane. For example, in the critically finite case, if all singular orbits escape, we cannot have any stable domains (recall that domains at ∞ are not possible for maps in the class S) and so $J = \mathbf{C}$.

Example. Let $E_\lambda(x) = \lambda e^x$ with $\lambda > 1/e$. Then the graph of E_λ (see Figure 3) shows that the orbit of 0 (the only singular value) escapes.

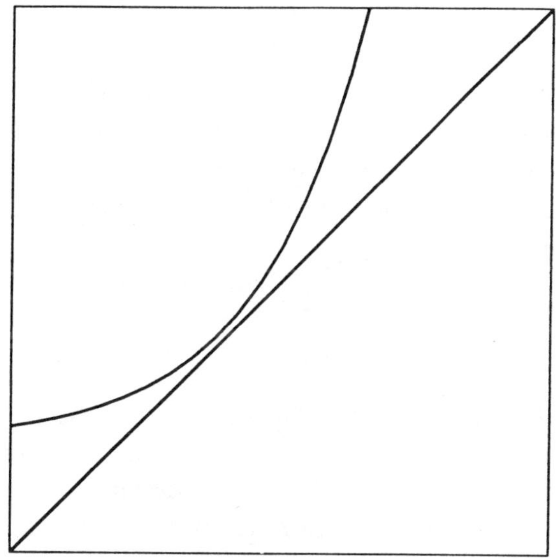

Fig. 3 The graph of $E_\lambda(x) = \lambda e^x$, $\lambda > 1/e$.

This example proves that $J(E_\lambda) = \mathbf{C}$. For $\lambda = 1$, this was conjectured by Fatou in [F] and first proved by Misiurewicz [Mi] some 60 years later using techniques that pre-dated the Sullivan classification above.

Example. $J(k\pi i e^z) = \mathbf{C}$ for any $k \in \mathbf{Z} - \{0\}$ since the orbit of 0 eventually lands on a repelling fixed point. $J(2k\pi \sin z) = \mathbf{C}$ for similar reasons.

4. Cantor Bouquets.

4. Cantor Bouquets. For quadratic polynomials, there are basically two types of Julia sets: Cantor sets and Julia sets that are connected. For critically finite entire functions there is a similar dichotomy: either the Julia set is the entire plane or else it contains Cantor bouquets. In this section we will sketch the construction of a simple Cantor bouquet.

Let $E_\lambda(z) = \lambda e^z$ with $0 < \lambda < 1/e$. The graph of E_λ (see Figure 1) shows that E_λ has two fixed points on the real axis, an attracting fixed point at q_λ and a repelling fixed point at p_λ, with $0 < q_\lambda < p_\lambda$. Note that $q_\lambda < -\log\lambda < p_\lambda$ since $E_\lambda'(q_\lambda) < E_\lambda'(-\log\lambda) = 1 < E_\lambda'(p_\lambda)$. Fix a real number ℓ_λ satisfying $-\log\lambda < \ell_\lambda < p_\lambda$ and observe that, if Re $z \geq \ell_\lambda$, then

$$|E_\lambda'(z)| = \lambda e^{\mathrm{Re}\,z} \geq \lambda e^{\ell_\lambda} > \mu > 1$$

for some constant $\mu > 1$.

Note that E_λ maps the half plane Re $z < \ell_\lambda$ inside itself, since $E_\lambda(\ell_\lambda) < \ell_\lambda$. Now E_λ has a fixed point in this half plane, namely q_λ. It follows from the Schwarz lemma that all orbits in this half plane tend to q_λ, and so the Julia set of E_λ is contained in the right-half plane Re $z \geq \ell_\lambda$.

We will now construct a collection of invariant Cantor sets for E_λ in the right-half plane Re $z \geq \ell_\lambda$ on which the dynamics of E_λ is conjugate to the one-sided shift map on $2N + 1$ symbols. Fix an integer $N \geq 1$. Consider the rectangle B_N bounded as follows:

1. on the left by Re $z = \ell_\lambda$
2. above and below by Im $z = \pm(2N + 1)\pi$
3. on the right by Re $z = r_\lambda$ where r_λ satisfies $\lambda e^{r_\lambda} > r_\lambda + (2N + 1)\pi$

The inequality in part 3 guarantees that E_λ maps the right-hand edge of the rectangle to a circle of radius λe^{r_λ} centered at 0 which contains the entire rectangle B_N in its interior. Note also that E_λ maps B_N onto the annular region $\{z \mid \lambda e^{\ell_\lambda} \leq |z| \leq \lambda e^{r_\lambda}\}$ and that B_N is contained in the interior of this annulus.

For each integer i with $-N \leq i \leq N$, consider the subrectangle $R_i \subset B_N$ defined by $\ell_\lambda \leq$ Re $z \leq r_\lambda$ and $(2i - 1)\pi \leq$ Im $z \leq (2i + 1)\pi$. Note that E_λ maps each R_i onto the annular region above, and that $|E_\lambda'(z)| > \mu > 1$. Moreover, if we restrict E_λ to the interior of R_i we obtain an expanding analytic isomorphism which maps the interior of R_i onto a region that covers all of B_N. See Figure 4. As a consequence, we may define an analytic branch of the inverse of E_λ, $L_{\lambda,i}$, which takes B_N into R_i for each i. Clearly, $L_{\lambda,i}$ is a contraction for each i.

Now define

$$\Lambda_N = \{z \in B_N | E_\lambda^j(z) \in B_N \text{ for } j = 0, 1, 2, \ldots\}.$$

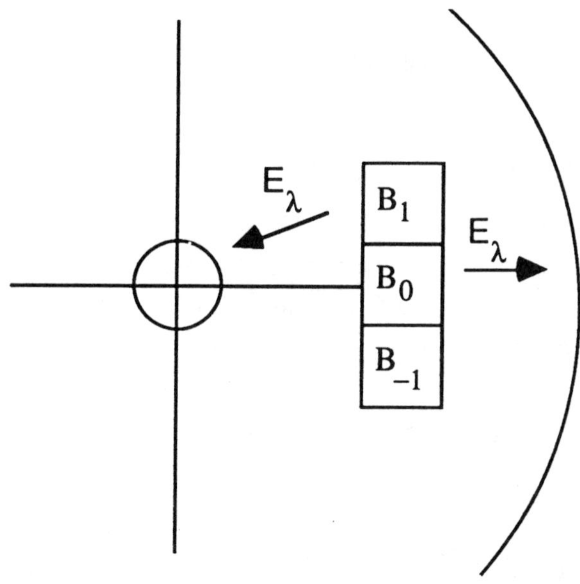

Fig. 4 Construction of B_N and R_i.

Let Σ_N denote the space of infinite sequences $\underline{s} = (s_0 s_1 s_2 \ldots)$ where each s_j is an integer, $-N \le s_j \le N$. Endow Σ_N with the product topology. For each $\underline{s} \in \Sigma_N$, we identify a unique point in Λ_N via

$$\phi(\underline{s}) = \lim_{n \to \infty} L_{\lambda, s_0} \circ L_{\lambda, s_1} \circ \cdots \circ L_{\lambda, s_n}(z)$$

where z is any point in B_N. In fact $\phi(\underline{s})$ is independent of z, and ϕ defines a homeomorphism between Σ_N and Λ_N. Moreover, ϕ gives a conjugacy between the shift map on the sequence space Σ_N and $E_\lambda | \Lambda_N$.

Since $|E_\lambda'(z)| > \mu > 1$ for all $z \in \Lambda_N$, it follows that Λ_N is a hyperbolic set, and so $\Lambda_N \subset J(E_\lambda)$. Moreover, we have an increasing sequence of these Cantor sets $\Lambda_1 \subset \Lambda_2 \subset \ldots$.

We next claim that each point $z_{\underline{s}} = \phi(\underline{s})$ in one of the Λ_i comes with a unique "hair" attached. This hair is a curve associated with a natural parametrization

$$h_{\underline{s}} \colon [1, \infty) \to \mathbf{C}$$

that satisfies

1. $h_{\underline{s}}(1) = z_{\underline{s}}$
2. $h_{\underline{s}}$ is a homeomorphism
3. If $t \neq 1$, then Re $E_\lambda^n(h_{\underline{s}}(t)) \to \infty$ as $n \to \infty$
4. For each t, $h_{\underline{s}}(t)$ lies in the horizontal strip

$$(2s_0 - 1)\pi < \text{Im } z < (2s_0 + 1)\pi$$

5. $E_\lambda(h_{\underline{s}}(t)) = h_{\sigma(\underline{s})}(E_{1/e}(t))$ where $\sigma(\underline{s})$ denotes the shift applied to the sequence $\underline{s}$.

To define $h_{\underline{s}}$, observe that $E_{1/e}$ has a unique fixed point at 1 and that $E_{1/e} : [1, \infty) \to [1, \infty)$. If $t > 1$, then $E_{1/e}^n(t) \to \infty$ as $n \to \infty$. Let L_{λ, s_i} denote the branch of the inverse of E_λ which now takes values in the horizontal strip given by

$$(2s_i - 1)\pi \leq \text{Im } z \leq (2s_i + 1)\pi.$$

Then define

$$h_{\underline{s}}(t) = \lim_{n \to \infty} L_{\lambda, s_0} \circ \ldots \circ L_{\lambda, s_{n-1}}(E_{1/e}^n(t)).$$

It can be shown that $h_{\underline{s}}$ has all of the properties listed above. Moreover, we may define

$$H_N : \Sigma_N \times [1, \infty) \to \mathbf{C}$$

by

$$H_N(\underline{s}, t) = h_{\underline{s}}(t).$$

The mapping H_N is a homeomorphism that takes $\{\underline{s}\} \times [1, \infty)$ onto the hair through $z_{\underline{s}}$ in Λ_N. Moreover, E_λ permutes these hairs just as the shift map permutes the sequences in Σ_N. Let Γ_N denote the image of $\Sigma_N \times [1, \infty)$ under H_N. Γ_N is called a *Cantor N-bouquet*. Note that Γ_N is invariant under E_λ. Moreover, the orbit of any point of the form $H_N(\underline{s}, t)$ with $t > 1$ tends to ∞ in the right-half plane. Since $|(E_\lambda^n)'| \to \infty$ along such an orbit, it follows that these points lie in the Julia set of E_λ. The remaining points, $H_N(\underline{s}, 1) = \Lambda_N$ also lies in $J(E_\lambda)$ as we saw above, but these points have bounded orbits. We call Λ_N the *crown* of the Cantor bouquet.

Let Γ denote the closure of the union of the Γ_N. Γ is called a *Cantor bouquet*. We claim that $\Gamma = J(E_\lambda)$. Clearly $\Gamma \subset J(E_\lambda)$, since each $\Gamma_N \subset J(E_\lambda)$. Conversely, if $z \in J(E_\lambda)$, then z is a limit of repelling periodic points, by Baker's result. Each such repelling cycle must lie in Λ_N for some N, hence z lies in the closure of $\bigcup_N \Lambda_N$.

Remarks.

1. The construction above works for any exponential for which there exists an attracting or neutral periodic point. However, in the general case, some of the hairs in the Cantor bouquet may be attached to the same point in the crown.

2. McMullen [Mc] has shown that the Hausdorff dimension of the Cantor bouquet constructed above is 2 but its Lebesgue measure is zero.

3. Mayer [M] has shown that this Cantor bouquet has the following remarkable topological property. Let D denote the crown of the bouquet. Then, in the Riemann sphere, $D \cup \{\infty\}$ is a connected set, but D itself is totally disconnected! Thus $D \cup \{\infty\}$ gives an example of a connected set from which we may remove one point to get a totally disconnected set.

4. Building on Mayer's results, Aarts and Oversteegen [AO] have shown that there is essentially only one way to embed a Cantor bouquet in the plane.

To see how Cantor bouquets arise in other critically finite maps, suppose all singular points of F lie in some disk B_r of radius r centered at the origin. Consider the preimage $F^{-1}(\mathbf{C} - B_r)$ and let U be any component of this set. Now F maps U analytically onto $C - B_r$ without singular values, so F must be a universal covering. As such, F acts like an exponential map. If, in fact, U is disjoint from $\mathbf{C} - B_r$ and F has sufficient growth in U, it can be shown (see [DK]) that there is an invariant Cantor bouquet for F in U.

5. Structural Instability. Our goal in this section is to begin to describe the parameter plane for $E_\lambda(z) = \lambda e^z$. In order to underscore the complexity of the parameter plane, we first describe the dynamics of E_λ when $\lambda = e^{i\theta}$ for small $\theta > 0$. That is, we consider perturbations of e^z within the exponential family.

As we have seen, it is the orbit of 0 which determines much of the dynamics of E_λ. If E_λ has an attracting periodic orbit, 0 is attracted to this orbit. If $E_\lambda^n(0) \to \infty$ or if 0 is preperiodic, then $J(E_\lambda) = \mathbf{C}$. So let us consider the orbit of 0 for each $\lambda = e^{i\theta}$ with $\theta > 0$ and small. Intuitively, multiplication by $e^{i\theta}$ gives successive points on the orbit of 0 a slight counterclockwise twist after each application of the exponential function. So we expect the orbit of 0 to "climb" in the imaginary direction when $\theta \neq 0$. That being the case, it is entirely possible that, for certain θ's, the orbit of 0 will land on the line Im $z = \pi$. If this occurs, then the next iterate lands in the far left-half plane, and one more iterate carries 0 back extremely close to itself. See Fig. 5.

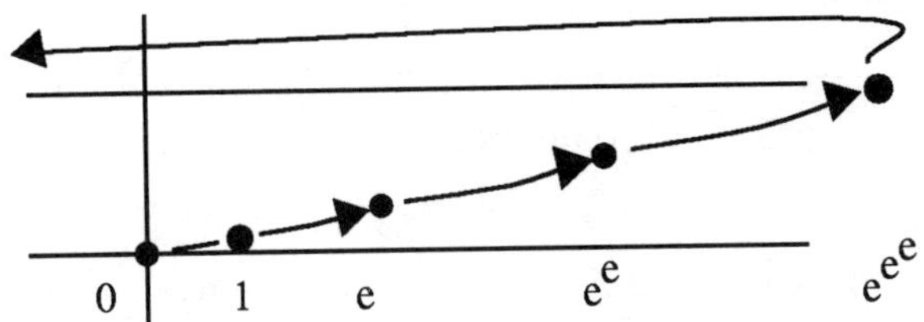

Fig. 5 The orbit of 0 under E_λ, $\lambda = e^{i\theta}$.

We emphasize that 0 comes back extremely close to itself, for the real parts of the orbit of 0 form a sequence which is approximately given by

$$0, 1, e, e^e, e^{e^e}, \ldots, E^{n-1}(e), -E^n(e), e^{-E^n(e)}$$

and, of course, $e^{-E^n(e)}$ is extremely small.

Also observe that it should be possible to select values θ_n with n sufficiently large so that 0 hops to the far left half plane precisely on iteration n.

Let us make all of this more precise. We will write $E_\lambda = E_\theta$, with $0 \leq \theta < \pi/2$. Let S be the strip determined by Re $z \geq 2$, $0 \leq$ Im $z \leq \pi$. Let $z_0 = x_0 + iy_0$ and write

$$z_1 = z_1(\theta) = x_1 + iy_1 = E_\theta(z_0).$$

Lemma. *There exists a θ_1 with $0 < \theta_1 < \pi/2$ such that, if $0 \leq \theta \leq \theta_1$ and both $z_0, z_1 \in S$, then*
    (i) $x_1(\theta) > 2x_0 + 1$
    (ii) $y_1(\theta) > 2y_0$.

Proof. First let $\theta = 0$. Then $e^{x_0} \sin y_0 \leq \pi$, so that $\sin y_0 \leq \pi e^{-2}$. Hence $\cos y_0 > 0.8$. Therefore,

$$x_1 = e^{x_0} \cos y_0 > 0.8 e^{x_0} > 2x_0 + 1.5.$$

Also,

$$y_1 = e^{x_0} \sin y_0 > e^{x_0} y_0/2 > y_0 e^2/2 > 3y_0.$$

So (i) and (ii) certainly hold for E_0. Now, for $0 < \theta < \pi/2$, note that $E_\theta(z) = e^{i\theta} E_0(z)$. If $z, E_\theta(z) \in S$, then it follows that $E_0(z) \in S$ as well. Hence it suffices to find θ_1 such that, if $\theta < \theta_1$, then

$$(1) \qquad 0 < \mathrm{Re}(w - e^{i\theta} w) < 1/2,$$

$$(2) \qquad \mathrm{Im}(e^{i\theta} w) > \mathrm{Im}(w)$$

for all $w \in S$ such that $e^{i\theta} w \in S$. Clearly (2) holds if $\theta_1 < \pi/2$. For (i), observe that if $e^{i\theta} w \in S$, then $\mathrm{Re}\, w < \pi/\sin\theta$. Therefore,

$$0 < \mathrm{Re}(w - e^{i\theta} w) < (\mathrm{Re}\ w)(1 - \cos\theta) + \pi\sin\theta$$

$$< \pi(1 - \cos\theta)/\sin\theta + \pi\sin\theta.$$

Since the right-hand side approaches 0 as $\theta \to 0$, we may choose θ_1 small enough so that (i) holds for $\theta < \theta_1$.

We remark that there exists $\theta_2 > 0$ such that $E_\theta^2(0) \in S$ for $0 \le \theta \le \theta_2$. From now on we assume that $\theta < \min(\theta_1, \theta_2)$.

Define $G_n(\theta) = E_\theta^n(0)$. Now $G_1(\theta)$ traces out the unit circle, while $G_2(\theta)$ gives a cardioid-like curve, part of which meets S. See Fig. 6. Let $\gamma_2(\theta)$ denote the piece of $G_2(\theta)$ in S. Note that $\gamma_2(\theta)$ meets $y = 0$ at $E_0^2(0) = e$. Let $\gamma_n(\theta)$ denote the connected component of $S \cap G_n(\theta)$ which contains $E_0^n(0)$. For n sufficiently large (numerically, $n \ge 3$), $\gamma_n(\theta)$ connects $y = 0$ to $y = \pi$ in S. More precisely, for $n \ge 3$, there exists θ_n such that $\mathrm{Im}\ (\gamma_n(\theta_n)) = \pi$ with $\mathrm{Re}\ (\gamma_n(\theta_n)) \ge 2$, and for all θ with $0 \le \theta < \theta_n$, $\gamma_n(\theta) \in S$. See Figure 6. This can be seen by applying the previous Lemma repeatedly to $E_\theta(\gamma_2(\theta))$. If $\theta > 0$, there exists $n = n(\theta)$ such that $E_\theta^n(\gamma_2(\theta)) \notin S$.

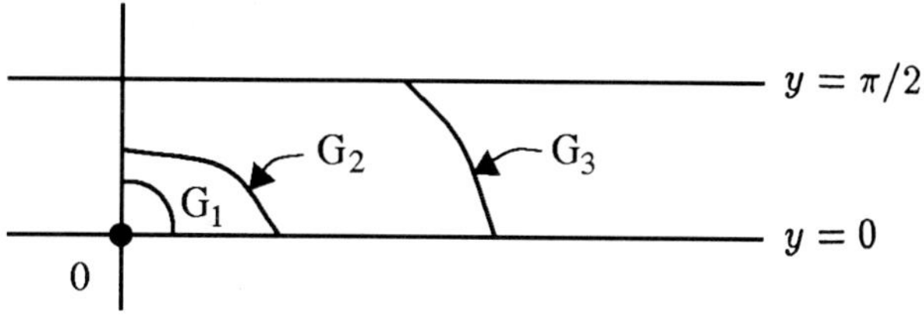

Fig. 6 The curves $G_n(\theta)$ for $n = 1, 2, 3$.

Clearly, $\exp(\gamma_n(\theta))$ is a curve in the upper half-plane which connects the positive real axis to the negative real axis. Since $0 < \theta < \pi/2$, the curve $E_\theta(\gamma_n(\theta)) = e^{i\theta}\exp(\gamma_n(\theta))$ also meets the negative real axis. When $\theta = \theta_n$, the image $E_\theta(\gamma_n(\theta))$ is negative and real. Denote this point by z_{n+1}, so that $z_{n+1} = E_\theta^{n+1}(0)$. Similarly let $z_j = E_\theta^j(0) = x_j + iy_j$ for $0 \le j \le n+1$. Note that $z_0 = 0, |z_1| = 1$, and $z_j \in S$ for $2 \le j \le n$.

Lemma. For fixed n, we have $\exp(x_n) \ge 2 + \sum_{j=1}^n (x_j + 1)$.

Proof. We have $x_2 \ge x_1 + 1$. Moreover, by the previous Lemma, for $2 \le j \le n-1$, we have $x_{j+1} \ge 2x_j + 1$. Hence,

$$2 + \sum_{j=1}^n (x_j + 1) \le \sum_{j=2}^{n-1}(x_{j+1} - x_j) + x_2 + x_n + 3 = 2x_n + 3 < e^{x_n},$$

since $x_n \ge 2$.

We now construct a disk about 0 which is contracted inside itself by E_θ^{n+2}. Let $r_{n+1} = 1$ and define $r_k = r_{k+1}/e^{x_k}e$ for $0 \le k \le n$. Note that $r_j < 1$ for $j \le n$ and $r_0 = (e^{n+1}\prod_{j=0}^n e^{x_j})^{-1}$. Let B_j be the disk of radius r_j about z_j.

Proposition. *If $z \in B_0$, then $E_\theta^j(z) \in B_j$ for $0 \le j \le n+1$. Moreover,*

$$\left|(E_\theta^{n+1})'(z)\right| \le e^{n+1}\prod_{j=0}^n e^{x_j}.$$

Proof. Suppose $|z - z_j| < r_j$. Let $M_j = \sup_{z \in B_j}|E_\theta'(z)|$. Then, for $j \le n$,

$$|E_\theta(z) - z_{j+1}| \le M_j r_j \le |\lambda e^{x_j + r_j}|r_j \le e^{x_j + r_j} \cdot r_j < e^{x_j}e \cdot r_j = r_{j+1},$$

since $r_j < 1$. Consequently, E_θ maps B_j strictly inside B_{j+1}, and we have $|E_\theta'(z)| \le e^{x_j+1}$. The result follows immediately.

Theorem. *E_θ has an attracting periodic point of period $n+2$ in B_0.*

Proof. By the preceding proposition, $E_\theta^{n+1}(B_0) \subset B_{n+1}$. We now show that $E_\theta(B_{n+1}) \subset B_0$. Let $z \in B_{n+1}$. Then $\operatorname{Re} z \le x_{n+1} + 1 = -e^{x_n} + 1$. Applying the previous Lemma, we have

$$|E_\theta(z)| \le \exp(-e^{x_n} + 1) \le e\exp\left(-\sum_{j=1}^n (x_j + 1) - 2\right)$$

$$= e^{-1}\left(\prod_{j=0}^n e^{x_j}\right)^{-1} e^{-n} = r_0,$$

as required. Hence, there exists a periodic point w such that $E_\theta^j(w) \in B_j$ for $0 \le j \le n+1$ and $E_\theta^{n+2}(w) = w$.

Finally, observe that $|E_\theta'(z)| < r_0$ if $z \in B_{n+1}$. Combined with the results of the Proposition, this yields $|(E_\theta^{n+2})'(w)| < 1$. This completes the proof.

Thus we have proved the following

Theorem. *There is a sequence $\theta_n \to 0$ such that, if n is sufficiently large, E_{θ_n} has an attracting periodic orbit of period n.*

There is nothing sacred about our using the parameter value $\lambda = 1$; one may check easily that the above arguments go through for any $\lambda > 1/e$ and parameter values $\lambda_\theta = \lambda e^{i\theta}$. The result is a sequence of open sets in the λ-plane converging to the interval $(1/e, \infty)$. For λ-values in these open sets, E_λ admits an attracting periodic orbit of period n.

As a corollary, we have

Corollary. *E_λ is not structurally stable for $\lambda > 1/e$.*

See [D1] for more details on the above results. As a remark, Douady and Goldberg [DoG] have proved that if $\lambda, \mu > 1/e$, then E_λ and E_μ are not topologically conjugate. This is true despite the fact that $J(E_\lambda) = J(E_\mu) = \mathbf{C}$.

There is much more to the bifurcation picture for E_λ when λ is near 1. Let us consider the functions

$$H_n(\lambda) = E_\lambda^n(0).$$

The map H_n is a function of the parameter λ and tells where the nth iterate of 0 lands for each λ. As a consequence of our previous work, the family of functions $\{H_n\}$ is not normal at any parameter value $\lambda > 1/e$. Fix $\lambda_0 > 1/e$ and let U be any neighborhood of λ_0 in the λ-plane. Then Montel's Theorem guarantees that the H_n assume all but 2 values in U infinitely often. These 2 omitted values are 0 and ∞, so, for example, there is $\lambda \in U$ such that $H_n(\lambda) = 2\pi i$. But then

$$E_\lambda^{n+1}(0) = E_\lambda(2\pi i) = \lambda = E_\lambda(0).$$

Consequently, for this λ-value, 0 is preperiodic, and so $J(E_\lambda) = \mathbf{C}$. It follows that every $\lambda > 1/e$ is a limit of points for which 0 is preperiodic.

Thus we have

Theorem. *If $\lambda > 1/e$, there is a sequence $\lambda_n \to \lambda$ such that 0 is preperiodic for E_{λ_n}, so that $J(E_{\lambda_n}) = \mathbf{C}$.*

We will discuss later the fact that each of these λ_n-values also has a "hair" attached in the λ-plane. This hair consists of λ-values for which $E_\lambda^n(0) \to \infty$. Consequently, $J(E_\lambda) = \mathbf{C}$ in this case, too.

6. The parameter plane for E_λ. Our goal in this section is to paint the picture of the parameter plane for the exponential family. We will see that there are some similarities with the Mandelbrot set, but a great many apparent dissimilarities. In the next section, we will see how to link these two pictures.

As in the case of the quadratic family, the parameter plane for E_λ is a picture of the fate of the critical orbit. In Fig. 7, we have displayed a portion of the parameter plane for E_λ. Black points indicate regions where the critical orbit is bounded, whereas white points indicate parameter values where the critical orbit escapes beyond a certain bound in the right half plane. These pictures demand some explanation. While it appears that there are large white regions of escape values, in fact the escape locus consists of a Cantor set of curves. This gives another instance of the parameter plane "resembling" the corresponding images in the dynamical plane, i.e., the Cantor bouquets discussed earlier. The reason for the large white regions is that the algorithm used to paint these pictures designates a point as escaping if its real part ever exceeds 50. The reason for this low choice of a bound is the problems encountered when computing e^{50}. Durkin [Du] gives a discussion of the accuracy and pitfalls of this algorithm.

Note that there is a large black cardioid-shaped region in the center of Fig. 7. As in the case of the Mandelbrot set, this is the set of parameter values for which E_λ has an attracting fixed point. To see this, note that E_λ has an attracting fixed point if

$$E_\lambda(z) = \lambda e^z = z$$
$$E_\lambda'(z) = \lambda e^z = \zeta$$

where $|\zeta| < 1$. These equations yield

$$\lambda = \zeta e^{-\zeta}, \ |\zeta| < 1.$$

This region is a cardioid C_1 in the plane as depicted in Fig. 8.

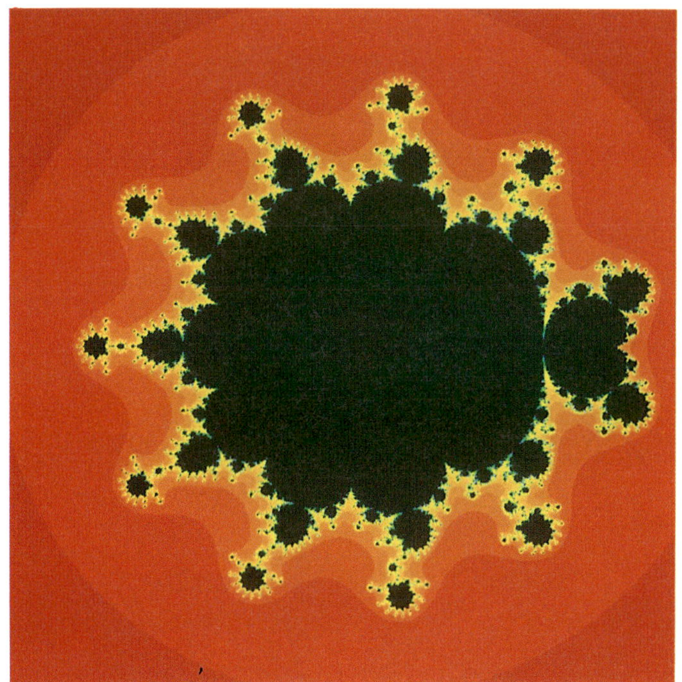

COLOR PLATE 1. B_d when $d = 10$.

COLOR PLATE 2. B_d when $d = 100$.

R. L. DEVANEY

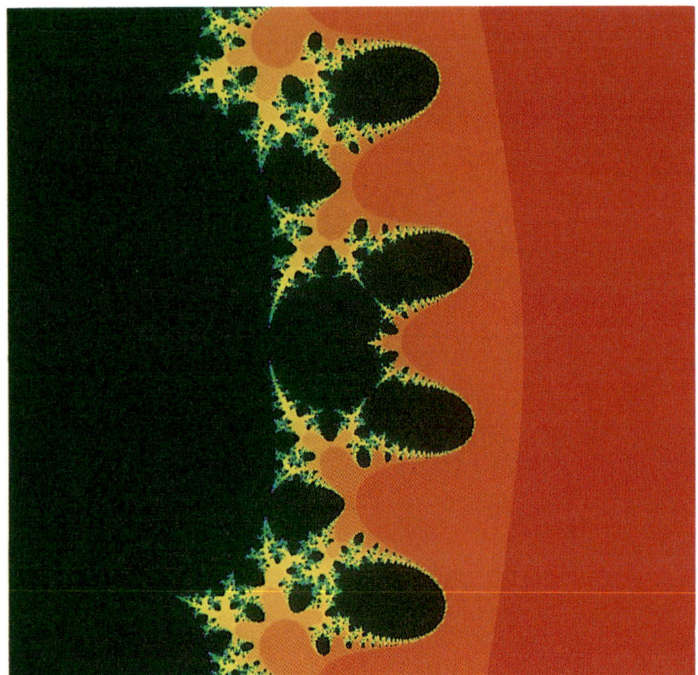

COLOR PLATE 3. The cardioid in B_{100}.

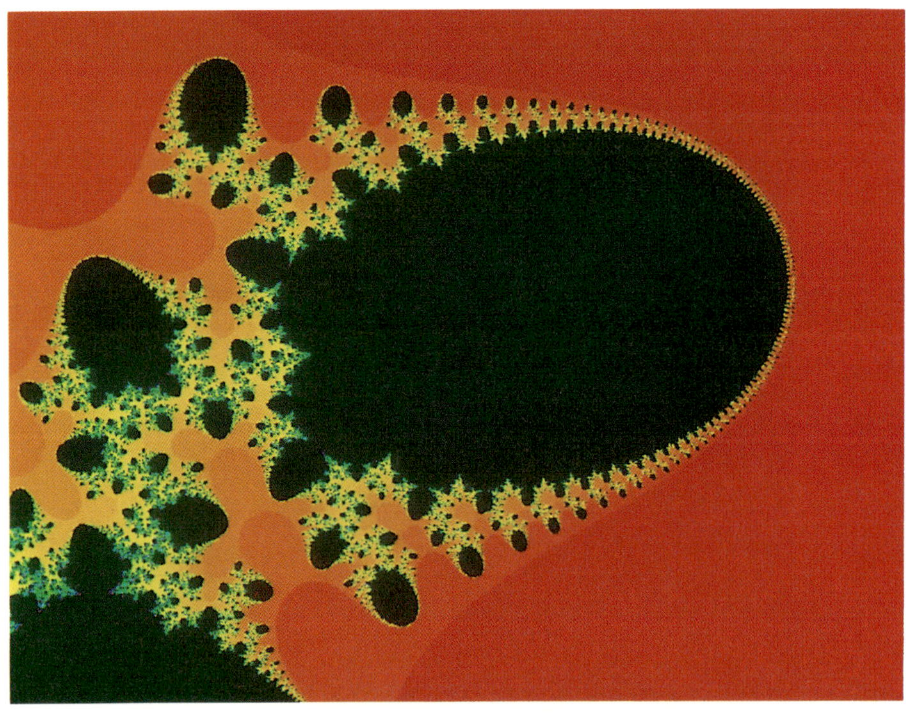

COLOR PLATE 4. An outlier in B_{100}.

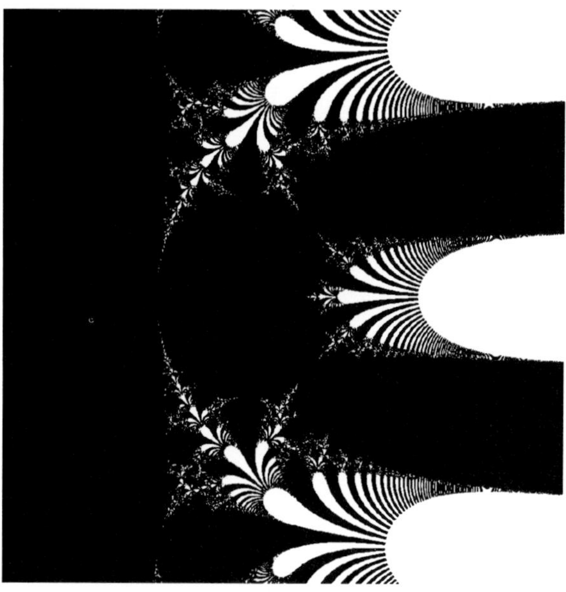

Fig. 7 The parameter plane for $E_\lambda(x) = \lambda e^z$.

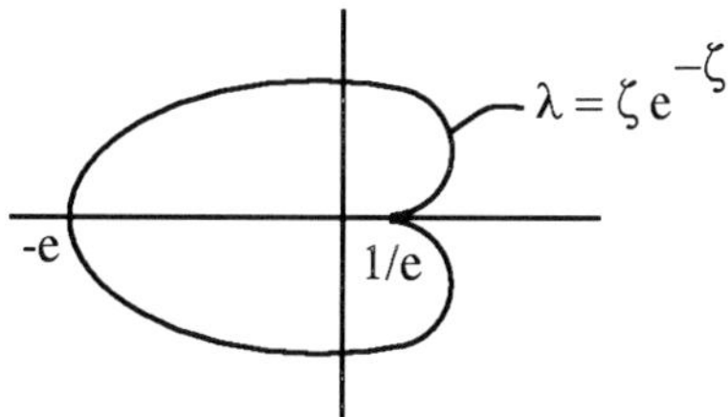

Fig. 8. The attracting fixed point region.

The cusp of the cardioid is at $\lambda = 1/e$, so our work in the previous section indicates that the structure of the parameter plane is quite rich near this point. There are open regions in which E_λ has attracting periodic points as well as preperiodic λ-values with hairs attached for which $J(E_\lambda) = \mathbf{C}$. See Fig. 9. The white regions in this image consist of λ-values for which the orbit of 0 under E_λ tends to ∞.

As λ traverses the boundary of the cardioid C_1, the multiplier ζ wraps once around the unit circle. Hence, we expect period-n bifurcations to occur whenever ζ is an nth root of unity. In the case of the Mandelbrot set, a similar picture occurs, but at these period-n bifurcation points, there is attached a small disk of parameter values corresponding to the attracting period-n orbit. In this case, our previous work suggests that the corresponding period n-regions are large, extending to ∞. See Fig. 10. This is indeed the case.

Suppose $z_0 = z_0(\lambda)$ is a periodic point for E_λ. We write $z_i =$

Fig. 9 The parameter plane near $\lambda > 1/e$.

Fig. 10. Period-n regions in the parameter plane.

$E_\lambda^i(z_0)$. Let C_k denote the set of λ-values for which E_λ has an attracting cycle of period k. To prove that each component of C_k is unbounded and simply connected, we first need a lemma.

Lemma. *Let $U \subset C_k$ be open with $\overline{U}$ compact and $0 \notin \overline{U}$. Then there exists constants c_1, c_2 such that, if $\lambda \in U$, then $c_1 \leq |z_j(\lambda)| \leq c_2$ for $j = 1, \ldots, k - 1$.*

Proof. Suppose some $|z_j(\lambda)| \to \infty$ as $\lambda \to \lambda_0$. For each $\lambda \in B_r(\lambda_0) \cap C_k$, there is at least one $z_i(\lambda)$ for which $|z_i(\lambda)| < 1$. This follows from the Chain Rule and the fact that $E_\lambda'(z_i) = z_{i+1}$ for all i. Then

$$c_2 = \max_{s=1,\ldots,k-1} |E_\lambda^s(z_i)|$$

is the required upper bound. Similar arguments give a lower bound.

Proposition. *Each component of C_k is simply connected.*

Proof. Let $G_n(\lambda) = E_\lambda^n(0)$. Observe that $\{G_n\}$ is a family of entire functions. Let C be a component of C_k; $\gamma \subset C$, a simple closed curve bounding a region D; and U, a small neighborhood of γ in C. We will show that $D \subset C$.

Since $U \subset C$, the functions $\{G_{nk}\}$ converge to the periodic point $z_0(\lambda)$ on U. By the Lemma, $|z_0(\lambda)| < c_2 < \infty$ on U, so it follows that the $\{G_{nk}\}$ converge on all of D, and the limit function determines a period k periodic point $z_0(\lambda)$ for all λ. By the maximum principle, $|\chi(\lambda)| < 1$ on D, where $\chi(\lambda)$ is the multiplier.

Proposition. *If $k \geq 2$, each component of C_k is unbounded.*

Proof. Let C be a compact component of C_k. By the lemma, $z_i(\lambda)$ is bounded away from 0 on C for all i, hence $\chi(\lambda)$ is as well. Therefore, if $\lambda_t \to \lambda \in \overline{C} - C$, then $|\chi(\lambda)| = 1$, and so the image of C under χ is closed in D.

Since χ is analytic on C, the image of C is also open. Hence $\chi(C) = D$, but $\chi(\lambda) \neq 0$ if $k \geq 2$. This contradiction establishes the result.

We remark that the period two region is unique in that it is the only period other than one for which there is a unique component in C_k. All larger periods have infinitely many components. The period two region is special since it occupies a large portion of the left half-plane. One may easily check that for any β, there is an α_0 such that, if $\alpha < \alpha_0$, then E_λ has an attracting cycle of period two if $\lambda = \alpha + i\beta$.

We remark that many of the above results were also proved by Baker and Rippon [BR]. The following result is due to John Hubbard.

Proposition. *Let χ denote the eigenvalue map on any component Q of C_k. If $k \geq 2$, then $\chi : Q \to D - \{0\}$ is a universal covering map.*

We refer to [DGH] for the proof. As a consequence of this, the map χ wraps around the unit circle infinitely often as λ traverses the boundary of Q. This is the reason for the cusps along the boundary of Q; at these parameter values, $\chi(\lambda) = 1$.

Now we turn to another phenomenon that appears in the parameter plane, that is, the λ-values for which $J(E_\lambda) = \mathbf{C}$. Most of these results appear in [DGH], so we just summarize them here. Recall that, in the previous section, we showed that, for each $\lambda > 1/e$, there is a limit point of λ-values for which 0 is preperiodic. Hence $J(E_\lambda) = \mathbf{C}$ for these values. We say that an itinerary $s = (s_0 s_1 s_2 \dots)$ is *regular* if $s_i \neq 0$ for any i. The following result is proved in [DGH].

Theorem. *Let s be a regular repeating itinerary. Then there is a unique $\lambda = \lambda_s$ such that λ_s has itinerary s under E_{λ_s}. Moreover λ_s is periodic, so that $J(E_{\lambda_s}) = \mathbf{C}$.*

We remark that the preperiodic λ-values produced in the previous section do not correspond to regular itineraries, so the above theorem is not optimal.

Theorem. *Let s be a regular repeating itinerary. Then there is a curve $\gamma_s(t)$ defined for $t \in [1, \infty)$ and satisfying*

    1. $\gamma_s(1) = \lambda_s$

    2. $\gamma_s(t)$ has itinerary s under $E_{\gamma_s(t)}$, and for each $t > 1$,

$$E_{\gamma_s(t)}^n(\gamma_s(t)) \to \infty$$

    as $n \to \infty$. Hence $J(E_{\gamma_s(t)}) = \mathbf{C}$.

    3. Re $\gamma_s(t) \to \infty$ as $t \to \infty$.

The curves $\gamma_s(t)$ provide the analogue of the hairs which occur in the dynamical plane.

We conjecture that, if the itinerary s is of the form

$$(s_1 \dots s_{n-1} 0 s_1 \dots s_{n-1} 0 \dots),$$

then the λ-values with this itinerary also form a curve in the λ-plane. We expect that this curve terminates at bifurcation points on the C_k rather than preperiodic points.

7. Relation with the Mandelbrot set.

At this juncture, there appears to be little similarity between the parameter plane for E_λ and the well-known Mandelbrot set, the parameter plane for the family of maps $z \mapsto z^2 + c$. Each family is a "natural" one-parameter family of maps with one singular point, and both parameter planes feature a central cardioid-like region in which the associated maps have an attracting fixed point. But there the similarity seems to end. In this section, however, we will show that there is a natural connection between the two sets.

The connection is given by the family of maps

$$P_{d,\lambda}(z) = \lambda \left(1 + \frac{z}{d}\right)^d.$$

Each $P_{d,\lambda}$ is a polynomial of degree d which has a single critical point at $-d$, and one critical value at 0. The family $P_{d,\lambda}$ provides the link between the quadratic family $z \mapsto z^2 + c$ and the exponential family. Let $Q_{d,c}(z) = z^d + c$ and define

$$v_c(z) = \frac{cz}{d} + c.$$

Then the reader may check easily that the affine map v_c conjugates $P_{d,\lambda}$ to $Q_{d,c}$, where c is any of the $d-1$ choices for which $\lambda = dc^{d-1}$.

We may construct, for each d, the parameter plane for the family $Q_{d,c}$. This proceeds in exact analogy with the construction of the Mandelbrot set as in [DH]. This parameter plane is a $(d-1)$-fold cover of the parameter plane for the family $P_{d,\lambda}$ with the projection given by

$$\lambda = dc^{d-1}.$$

Let B_d denote the parameter plane for the $P_{d,\lambda}$. Then, arguing as in [DH], we have

Theorem. *B_d is connected and the complement of B_d in $\overline{\mathbf{C}}$ is an open disk.*

If $\lambda \in B_d$, then it is known that the filled-in Julia set of $P_{d,\lambda}$ is connected, whereas, if $\lambda \in \mathbf{C} - B_d$, then

$$P_{d,\lambda}^n(0) \to \infty$$

and so $J(P_{d,\lambda})$ is a Cantor set. See [B].

In Fig. 11, we have displayed B_d for $d = 2, 3, 4,$ and 8. Note how the number of cusps on the "decorations" attached to the basic cardioid grows with d. Indeed, as $d \to \infty$, these figures "converge" to the parameter plane of E_λ. This is the link between the quadratic and exponential families.

Note that there is a fundamental difference between the B_d and the parameter plane for E_λ which we call B_∞. For each d, B_d is compact and its exterior is isomorphic to the unit disk. On the other hand, as we showed earlier, each component of C_k, $k \geq 2$, in B_∞ is non-compact, and the "exterior" of B_∞ contains the hairs described above. (Actually, the exterior of B_∞ is not defined since any hair in the λ-plane is a limit of components of the C_k, in exact analogy with what we showed in section 5.)

One question is what happens to the external rays of Douady and Hubbard which foliate the disk in the complement of B_d. In [DGH],

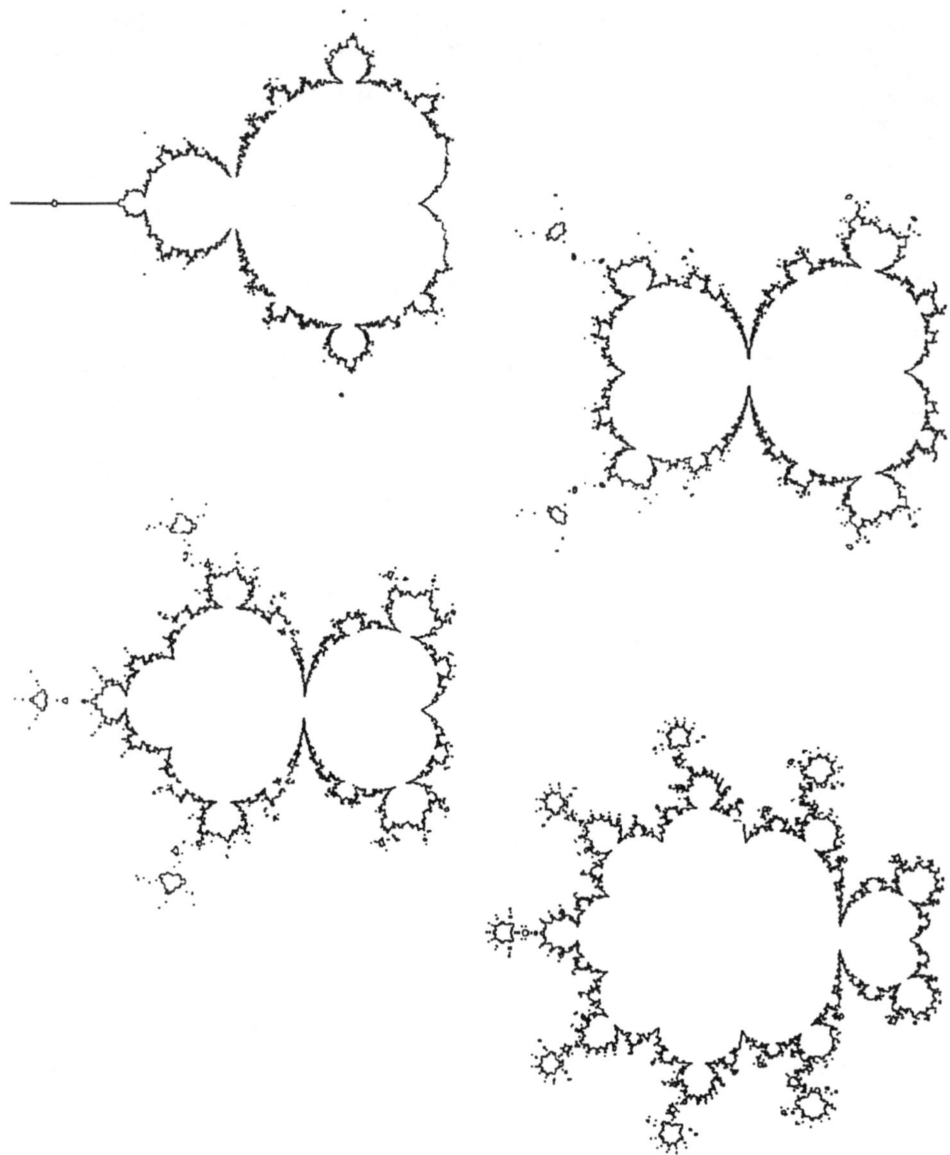

Fig. 11. B_d for $d = 2, 3, 4, 8$.

it is shown that certain of these rays have a limit as $d \to \infty$, and that
this limit is precisely one of the hairs in B_∞.

How do we identify the hairs which are limit curves of the external rays? The precise answer is spelled out in [DGH], but we will
give some special cases here. The external arguments of Douady and
Hubbard are given by rational numbers between 0 and 1. For example, for each d, the external argument of angle $1/d$ converges to a
preperiodic λ-value in B_d which is attached to the first "bulb" of the

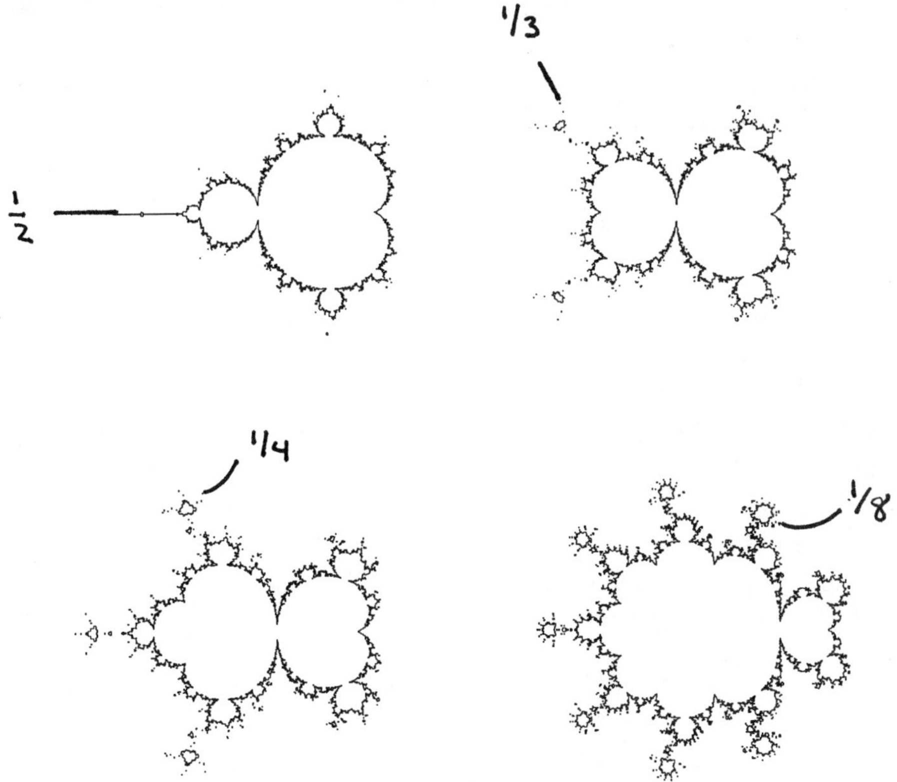

Fig. 12 The external argument corresponding to
angle $1/d$ for $d = 2, 3, 4, 8$.

large period 2 region. These angles are depicted in Fig. 12.

Note that, in base d, we may write

$$\frac{1}{d} = (d-1) \sum_{i=1}^{\infty} \frac{1}{d^{i+1}}.$$

The result in [DGH] is that this ray converges as $\lambda \to \infty$ to the hair on which λ has itinerary $(111...)$ (This is the hair which terminates at 2π.)

In general, if for each d, $p(d)$ is the angle given by

$$p(d) = (d-1) \sum_{i=1}^{\infty} \frac{s_i}{d^{i+1}}$$

where the s_i form a repeating, regular sequence, then the rays with angle $p(d)$ in B_d converge as $d \to \infty$ to the hair on which all λ's have itinerary $(s_1 s_2 s_3 \ldots)$.

It is an open question as to just how generally this result holds. For example, we conjecture that if $\mathbf{s} = (s_0 s_1 s_2 \ldots)$ is a repeating sequence with $s_0 = 0$, then the corresponding hair converges to a bifurcation point in the parameter plane with external angle

$$p(d) = (d-1) \sum_{i=0}^{\infty} \frac{s_i}{d^{i+1}}.$$

In particular, the hairs belonging to itineraries of the form $(0j0j0j\ldots)$ should have external angle $j/(d+1)$. See Fig. 13.

In Plates 1 and 2 we have provided color images of B_d for $d = 10$ and $d = 100$. The large circular region in B_{100} is the period-2 regime. Note how large it is relative to the case $d = 10$. At the far right hand side of this image is the cardioid which is magnified in Plate 3. The cardioids in B_{10} and B_{100} are essentially the same size. Plate 4 depicts one of the "outliers" when $d = 100$.

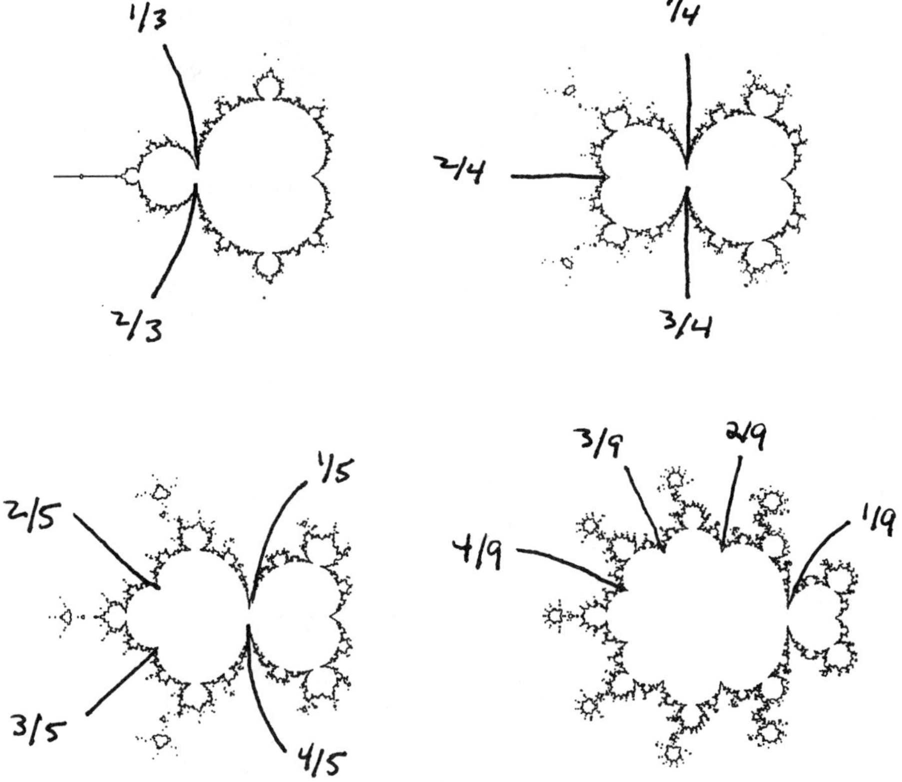

Fig. 13 The external argument belonging to angle $j/(d+1)$ for $d = 2, 3, 4, 8$.

References

[AO] Aarts, J. and Oversteegen, L. A Characterization of Smooth Cantor Bouquets. Preprint.

[Ba] Baker, I. N. Wandering Domains in the Iteration of Entire Functions, *Proc. London Math. Soc.* **49** (1984), 563-576.

[BR] Baker, I. N. and Rippon, P. Iteration of Exponential Functions, *Ann. Acad. Sci. Fenn.*, Series 1A Math. Vol. 9, 1984, pp. 49-77.

[B] Blanchard, P. Complex Analytic Dynamics on the Riemann Sphere, *Bull. AMS* Vol. II, No.1, 1984, 85-141.

[D1] Devaney, R. L. Structural Instability of Exp(z). *Proc. A.M.S.*, Vol. 94, 1985, pp. 545-548.

[D2] Devaney, R. L. Julia Sets and Bifurcation Diagrams for Exponential Maps. *Bull AMS* **11** (1984), 167-171.

[D3] Devaney, R. L. e^z: Dynamics and Bifurcations. *Intl. J. Bifurcation and Chaos.* **1** (1992), 287-308.

[DD] Devaney, R. L. and Durkin, M. The Exploding Exponential and Other Chaotic Bursts in Complex Dynamics. *American Mathematical Monthly* **98** (1991), 217-233.

[DG] Devaney, R. L. and Goldberg, L. Uniformization of Attracting Basins for Exponential Maps. *Duke Mathematics Journal.* **55** (1987), 253-266.

[DGH] Devaney, R. L., Goldberg, L., and Hubbard, J. A Dynamical Approximation to the Exponential Map by Polynomials. Preprint.

[DH] Douady, A. and Hubbard, J. Étude Dynamique des Polynôme Complexes, *Publications Mathematiques d'Orsay*, 84-102.

[DKeen] Devaney, R. L. and Keen, L., eds. *Chaos and Fractals: The Mathematics Behind the Computer Graphics* American Mathematical Society, Providence, 1989.

[DK] Devaney, R. L. and Krych, M. Dynamics of Exp(z), *Ergodic Theory and Dynamical Systems* **4** (1984), 35-52.

[DT] Devaney, R. L. and Tangerman, F. Dynamics of Entire Functions Near the Essential Singularity, *Ergodic Theory and Dynamical Systems* **6** (1986), 489-503.

[DoG] Douady, A. and Goldberg, L. The Nonconjugacy of Certain Exponential Functions. In *Holomorphic Functions and Moduli I*. MSRI Publ., Springer Verlag (1988), 1-8.

[Du] Durkin, M. B. The Accuracy of Computer Algorithms in Dynamical Systems. *Intl. J. Bifurcation and Chaos*. **1** (1991), 625-640.

[Er] Eremenko, A. On the Iteration of Entire Functions. In *Dyn. Syst. and Erg. Theory* Banach Center Publications. **23** (1989), 339-345.

[EL] Eremenko, A. and Lyubich, M. Yu. Iterates of Entire Functions. *Dokl. Akad. Nauk SSSR* **279** (1984), 25-27. English translation in *Sov. Math. Dokl.* **30** (1984), 592-594.

[EL 1] Eremenko, A. and Lyubich, M. Yu. Structural stability in some families of entire functions. *Funk. Anal. i Prilo.* **19** (1985), 86-87.

[F] Fatou, P., Sur l'Itération des fonctions transcendentes Entières, *Acta Math.* **47** (1926), 337-370.

[GK] Goldberg, L. R. and Keen, L. A Finiteness Theorem For A Dynamical Class of Entire Functions, *Ergodic Theory and Dynamical Systems* **6** (1986), 183-192.

[L] Lyubich, M. Measurable Dynamics of the Exponential, *Sov. Math. Dokl.* **35** (1987), 223-226.

[M] Mayer, J. An Explosion Point for the Set of Endpoints of the Julia Set of $\lambda \exp(z)$. *Ergodic Theory and Dynamical Systems* **10** (1990), 177-184.

[Mc] McMullen, C. Area and Hausdorff Dimension of Julia Sets of Entire Functions. *Trans. A.M.S.* **300** (1987), 329-342.

[Mi] Misiurewicz, M., On Iterates of e^z, *Ergodic Theory and Dynamical Systems* **1** (1981), 103-106.

[Ye] Ye, Z. Structural Instability of Exponential Functions. To appear in *Trans. A.M.S.*

Index

Other Titles in This Series

23 **R. V. Hogg, editor,** Modern statistics: Methods and applications (San Antonio, Texas, January 1980)

22 **G. H. Golub and J. Oliger, editors,** Numerical analysis (Atlanta, Georgia, January 1978)

21 **P. D. Lax, editor,** Mathematical aspects of production and distribution of energy (San Antonio, Texas, January 1976)

20 **J. P. LaSalle, editor,** The influence of computing on mathematical research and education (University of Montana, August 1973)

19 **J. T. Schwartz, editor,** Mathematical aspects of computer science (New York City, April 1966)

18 **H. Grad, editor,** Magneto-fluid and plasma dynamics (New York City, April 1965)

17 **R. Finn, editor,** Applications of nonlinear partial differential equations in mathematical physics (New York City, April 1964)

16 **R. Bellman, editor,** Stochastic processes in mathematical physics and engineering (New York City, April 1963)

15 **N. C. Metropolis, A. H. Taub, J. Todd, and C. B. Tompkins, editors,** Experimental arithmetic, high speed computing, and mathematics (Atlantic City and Chicago, April 1962)

14 **R. Bellman, editor,** Mathematical problems in the biological sciences (New York City, April 1961)

13 **R. Bellman, G. Birkhoff, and C. C. Lin, editors,** Hydrodynamic instability (New York City, April 1960)

12 **R. Jakobson, editor,** Structure of language and its mathematical aspects (New York City, April 1960)

11 **G. Birkhoff and E. P. Wigner, editors,** Nuclear reactor theory (New York City, April 1959)

10 **R. Bellman and M. Hall, Jr., editors,** Combinatorial analysis (New York University, April 1957)

9 **G. Birkhoff and R. E. Langer, editors,** Orbit theory (Columbia University, April 1958)

8 **L. M. Graves, editor,** Calculus of variations and its applications (University of Chicago, April 1956)

7 **L. A. MacColl, editor,** Applied probability (Polytechnic Institute of Brooklyn, April 1955)

6 **J. H. Curtiss, editor,** Numerical analysis (Santa Monica City College, August 1953)

5 **A. E. Heins, editor,** Wave motion and vibration theory (Carnegie Institute of Technology, June 1952)

4 **M. H. Martin, editor,** Fluid dynamics (University of Maryland, June 1951)

3 **R. V. Churchill, editor,** Elasticity (University of Michigan, June 1949)

2 **A. H. Taub, editor,** Electromagnetic theory (Massachusetts Institute of Technology, July 1948)

1 **E. Reissner, editor,** Non-linear problems in mechanics of continua (Brown University, August 1947)

(See the AMS catalog for earlier titles)

ISBN 0-8218-0290-9